ENVIRONMENTAL SCIENCE ACTIVITIES

Dorothy B. Rosenthal
Science Education Consultant

John Wiley & Sons, Inc.
New York • Chichester • Brisbane • Toronto • Singapore

ISBN 0-471-07626-0

Printed in the United States of America

10 9 8 7 6 5 4 3 2 1

Preface

Remark the intricacy of the skein, the complexity of the web.

Marcus Aurelius
A.D. 121-180

These words of Marcus Aurelius, written almost 2,000 years ago, are the essence of modern environmental science. All of the discoveries made since his time have only underscored his perceptions and given us greater appreciation for his insight. My purpose in writing *Environmental Science Activities* is to help you to discover for yourself the intricate and complex nature of the web of life. Whether you are enrolled in a course or investigating the environment on your own, these activities will help you to appreciate the interrelationships of all living things to each other and to their environment.

Today, however, appreciation and enjoyment of the environment are not enough. Human activities are creating environmental problems that affect everyone, and place all of us in a decision-making role about environmental issues. To make intelligent decisions, we need to understand how the environment works; therefore, each of the activities in this book was designed to teach one or more basic concepts of environmental science.

Environmental issues generate strong opinions, extreme positions, and contradictory claims. To analyze these issues requires not only understanding of the environment, but also an ability to evaluate evidence and arguments critically. Therefore, another major purpose of this book is to provide opportunities for you to practice critical thinking and scientific methods of inquiry. I have intentionally selected open-ended, investigative activities that ask you to make decisions about methods, develop hypotheses, vary experimental conditions, select subjects for field studies, evaluate evidence, participate in simulations, apply analogical thinking, propose alternative interpretations of data, or draw conclusions.

Because some of you will be doing these activities on your own, I have designed them so they can be done almost entirely with readily-available materials. A list of sources for those materials that are more difficult to obtain appears at the end of the book.

As a result of carrying out these activities, you may want to pursue study of the environment in greater depth, become actively involved in the environmental movement, or become a more environmentally-aware shopper and citizen. You may even want to go on to do research in environmental science. Most important of all, from my point of view, is for you to develop a lifelong interest in, and appreciation of, the environment.

I wish to thank Dr. Daniel B. Botkin, Professor, George Mason University, Fairfax, Virginia, and Director, Center for the Study of the Environment, Santa Barbara, California, for suggestions at many stages of this project. Dr. Allan Miller, Professor, California State University, Long Beach, California, reviewed an early draft of the manuscript and made many helpful comments. Kim Anna Blakemore, George Mason University, provided valuable research assistance. Erica Liu, Assistant Editor at John Wiley & Sons, Inc., has been a gracious and efficient facilitator throughout. Most of all, I owe a grateful and heartfelt thanks to my husband, Jerome A. Rosenthal, for providing the graphics, computer assistance, and loving support.

Dorothy B. Rosenthal
Indian Lake, New York

About the Author

Dorothy B. Rosenthal, a Science Education Consultant, has developed curriculum materials and contributed to textbooks in biology and environmental science. She has worked with the California and New York State Departments of Education on curriculum reform projects. From 1988 to 1993 she was on the faculty of the Science Education Program at California State University, Long Beach. Prior to that, she spent fifteen years as a secondary biology teacher and science supervisor in Henrietta, New York. Her articles have appeared in numerous science education publications.

How to Use This Book

Many of the activities in this book are designed to be done out-of-doors. These are indicated by a

next to the activity number.

Each activity begins with background information, which is set off by bold, horizontal lines from the rest of the text. This introduction is meant to complement a textbook; you will gain a deeper understanding of environmental science if you read related sections in a textbook as you do each activity.

A ☞ below the introduction points the way for you; it tells you briefly what you will be doing in the activity.

A ✔ appears next to a list of materials for most activities. It is assumed that you will have a basic tool kit available for all activities: pencil, pen, notebook, unlined paper, and ruler. These are **not** listed; therefore, if an activity requires no additional materials, you will not find a list of materials.

Most of the activities in this booklet can be completed using common materials that are available in grocery, variety, or hardware stores. Suggestions about how to obtain other materials appear at the end of the book in a section called "Sources of Information and Materials."

An estimate of the amount of time required to complete each activity is indicated by a ☺. This is, however, only an estimate and you will need to evaluate the time requirements for your own situation. Although most of the activities can be done in a few hours, many of them could well be extended into long-term projects.

A ❀ symbol introduces the instructions for each activity. It is a good idea to read all of these before beginning the activity. Planning and organization will minimize wasted time and effort.

The last section of each activity is the "Followup," indicated by a . This section gives you an opportunity to reflect on what you have done and learned, and to reinforce the concept(s) related to the activity.

indicates alternative activities. This means that you can choose between activities related to a basic concept.

Some alternative activities require a microscope. These have a symbol in front of them.

A few activities are similar in procedure to other activities in this book. These related activities are indicated by a double-headed arrow, as shown below:

You may want to do related activities at the same time to save you from collecting similar information more than once.

If you are interested in pursuing a topic in greater depth, you will find a list of references at the end of the book.

Table of Contents

INTRODUCTION

Activity #1: People and Nature

The relationship between humans and nature has changed considerably from the time when all people lived within a natural setting until the present, when most of us live in an environment that has been greatly affected by human activities. While the number of people on Earth was relatively small, its resources seemed endless; and the wilderness was something to be tamed and conquered. As society became more urban, people became cut off from nature. Parks, zoos, national parks, nature preserves, and wilderness areas were set aside to provide ways for people to interact with nature. Now that the human population is threatening to overwhelm the planet and to exhaust its resources, people are beginning to view nature as an ally and trying to find ways of living more harmoniously with the natural world. One way to begin your study of the environment is to observe the interactions between people and nature in a place set aside for that purpose.

☞In this activity, you will observe interactions between people and nature in a public park.

☉**TIME:**   2 hours for activity, 1 hour for followup

❀**ACTIVITY:**

1.   Select a park or other area that has been set aside for the public to enjoy nature. Obtain copies of literature about the park, if available. Visit the park on a day and at a time when it is likely to be busy. If the park is very large, select one area to study, preferably one that is well-used. Record observations of interactions between people and nature, using the questions below as a guide:

▸   How many people did you see?
▸   What types of activities did you observe them doing?
▸   What attitudes toward nature were evident? (see box on page 4)

2.   Record your impressions of the park on the "Park Profile Form" (see page 5).

3.   When you have completed the form, select one of the interactions or impressions and collect quantitative data related to it. For example, you may want to record how much time people spend watching animals. If the park seems to suffer from considerable abuse, you can count the number of examples of vandalism or the amount of litter. When you have selected the data you wish to collect, prepare a data table and record your observations.

✎FOLLOWUP:

1.   What is the goal of the park with respect to interactions between people and nature, and how is the park organized to achieve its goal? How could the park be improved for the benefit of people and the natural environment in the park?

2.   Compare the results of the data you collected in step #3 with your initial observations or impressions. What conclusions can you draw from your data?

3.   Pretend that you are writing an entry in a guide book about parks in your area. Using the information in your "Park Profile Form," write a brief description of the park, including its good and bad points, and an overall rating of the park on a scale of 1 to 5 (where 5 is outstanding).

4.   What attitudes toward nature did you see in the park? on the part of those who designed the park? on the part of those using the park?

Attitudes Toward Nature

Look for evidence of these attitudes toward nature.

1. Nature is something separate from people.
2. Nature is important mainly for its economic value.
3. Nature is something to be subdued and controlled by technology.
4. Nature is something to be studied scientifically.
5. Nature is an ideal condition to which we should return.
6. Nature knows best.
7. People should act as stewards of nature.
8. Nature is something to be enjoyed for its beauty.
9. There is unity in nature; each life form is part of the web of life.
10. Nature is wild, unpredictable, and frightening.
11. Nature is awesome and wondrous.

Park Profile Form

Rate the park on each of the items below, using the following scale:
5 - outstanding; 4 - excellent; 3 - good; 2 - fair, 1 - poor.

Park Characteristic	5	4	3	2	1
Amount of traffic	☐	☐	☐	☐	☐
Noise level	☐	☐	☐	☐	☐
Amount of litter	☐	☐	☐	☐	☐
Condition of facilities (sidewalks, benches, etc.)	☐	☐	☐	☐	☐
Amount of abuse (graffiti, vandalism, etc.)	☐	☐	☐	☐	☐
Pollution level	☐	☐	☐	☐	☐
Condition of ground (erosion of soil or grass, hard packing of ground, trampled on vegetation, damage beyond marked trails)	☐	☐	☐	☐	☐
Amount and condition of vegetation	☐	☐	☐	☐	☐
Amount of cover for animals	☐	☐	☐	☐	☐
Provisions for people	☐	☐	☐	☐	☐

Activity #2: Earth Systems

Systems are interacting collections of subunits with inputs and outputs. Feedback occurs when some or all of the output is fed back to another part. If feedback decreases the output, it is called negative feedback; when it increases the output, it is called positive feedback. Negative feedback tends to stabilize conditions, as when the oxygen level of blood in humans is maintained during exercise by increases in heart and breathing rates. Positive feedback tends to destabilize a system, as when fever causes dehydration; loss of water results in higher temperature, which leads to further dehydration. In everyday life we often describe the latter type of situation as a "vicious circle."

The Earth can be thought of as a system comprised of four major subsystems: the solid Earth, the atmosphere, the waters, and the biosphere. Drastic or rapid changes in conditions on Earth, such as the average global temperature, can be disastrous for life. Scientists have long been puzzled by evidence that the sun's luminosity has increased by 25% since the time when the first organisms to leave a fossil record lived. Yet, the average global temperature has not increased to that extent. A reasonable conclusion is that negative feedback has changed the Earth system in ways that compensated for the increase in energy input from the sun. Scientists are currently debating whether this is a happy accident or, as some have suggested, the Earth is a self-regulating system. James Lovelock has developed a model, the Daisy World, as an example of how living organisms could participate in such a system.

☞In this activity, you will predict changes in the Daisy World model.

✔**MATERIALS:**   2 empty soup cans
2 thermometers
black construction paper or cloth
white construction paper or cloth
clear tape or rubber band
graph paper

☺**TIME:**   2 hours for activity, 1 hour for followup

❧ACTIVITY:

1.   Remove the labels from the cans and wash and dry the cans and lids. Punch a hole, large enough to just fit the thermometers, in the center of each lid. Wrap black paper or cloth tightly around one can and secure with a rubber band or tape. Cut a circle the size of the lid from the paper or cloth. Make a hole in the center of the circle and tape the circle to the lid. Use the white paper to cover the other can and lid in the same manner. Tape the lids to the cans and put the thermometers through the holes in the lids to the same height. Record the temperatures on the thermometers. Put the cans in a sunny area and record the temperatures every 5 minutes until the temperatures level off. Graph the results with temperature on the vertical axis and number of minutes on the horizontal axis. What can you conclude about dark- and light-colored objects?

2.   Study the description of the Daisy World (see box). From your observations in step #1, what effect would a large number of dark daisies have on the Daisy World? of light daisies? Explain.

The Daisy World Assumptions

1. The atmosphere is clear and water covers the entire surface.
2. There are two species of plants, the daisies, which can exist at temperatures between 5°C and 40°C.
3. One species is light-colored; the other is dark-colored.
4. The dark species grows best in cool conditions, with its maximum growth occurring at 22.5°C.
5. The light species grows better under warmer conditions.
6. Herbivores, which feed on the daisies, are grey.
7. There are no other living organisms.

---

3.   Prepare a second graph, with temperatures from 0°C to 40°C on the horizontal axis. The vertical axis will represent relative numbers of daisies. Draw a line on the graph representing the change in numbers of dark daisies as temperature increases from 0°C to 40° C. Do the same for the light daisies.

The Gaia Hypothesis

The idea that the Earth is a self-regulating entity analogous to an organism was named by its originators, James Lovelock and Lynn Margulis, for Gaia, the Greek goddess of the Earth. Gaia includes three basic ideas:

- Life affects the physical environment.
- Life helps to stabilize the environment.
- The Earth, including living organisms, is a self-regulating system that maintains conditions suitable for life.

Most scientists accept the first statement; some accept the second. But the idea of the Earth as self-regulating, in a manner analogous to an organism, is highly controversial. For the majority of scientists, the last statement is not a hypothesis, but a myth or a metaphor, because it cannot be tested. But, whatever it is, Gaia has been useful in focusing attention on self-regulating mechanisms in the biosphere and on the Earth as a system.

✎FOLLOWUP:

1.   Draw a diagram that illustrates how negative feedback operates in the Daisy World model.

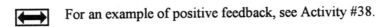   For an example of positive feedback, see Activity #38.

2.   In what ways are the assumptions of the Daisy World model realistic? unrealistic?

3.   Describe an example of negative feedback in the environment.

SCIENTIFIC METHODS

Activity #3: Bird Food

In the activity on "People and Nature," you were asked to make quantitative observations on one aspect of the park you were observing in order to check your preliminary impressions. Collecting quantitative data in as objective a manner as possible is one of the hallmarks of science. Often, the data bear out initial impressions; although such data do not add any new information, they do provide a firmer basis for accepting the original observations. In many instances, however, the quantitative data demonstrate that initial impressions are incorrect. As you learn about the environment through your activities or through the media, you need to evaluate the evidence for the information you receive. By carrying out your own investigations in this and the next activity, you will be in a better position to evaluate the data you gather in subsequent activities and the information you read in books, magazines, and newspapers and hear on radio and television.

☞In this activity, you will investigate the feeding preferences of birds as an example of the scientific approach to the study of the environment.

✔**MATERIALS:**   3 pieces of cardboard, 8 1/2" x 11"
watch or clock
3 types of bird seed; approximately 3 cups
        of each (see box on page 15)
measuring cup
field guide to common birds of your area
        (optional)

☺**TIME:**   1 hour for activity, 1 hour for followup

❀**ACTIVITY:**

1.   You will need to make observations for 5 minutes on 5 consecutive days at the same time of day (early morning or late afternoon are best), and at a location where birds are seen on a regular basis. If it rains on one day, skip that day and add another observation day to the end of the period.

2.   Place the three pieces of cardboard on the ground in a triangle, so that each piece is exactly 1 meter from the others. All three should be at least 1.5 meters from cover, such as trees, bushes, or tall grass. The cardboards should be at least 3 meters from any existing bird feeders. If it is windy, place a rock or other weight on each piece of cardboard.

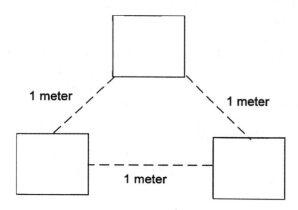

3.   Measure 1/2 cup of one type of food and place it in the middle of one of the pieces of cardboard. Do the same with the other types of food.

4.   Wait until a bird takes food from one piece of cardboard. For the next 5 minutes, record the numbers of visits made to each type of food by each type of bird. You can identify the birds from the field guide, or you can use a descriptor of a few words to identify each type.

5.   At the end of the 5 minutes, collect the foods and cardboard. When you make your next observation, place the cardboard triangle in the same location.

6.   Following are some things for you to consider to make your study as objective as possible. What constitutes a visit? (You will need to define it carefully and apply the definition consistently.) What if the position of the food in the triangle influences a bird's choice? (You will need to find a way to eliminate this possibility.) Does your presence make a difference? (How can you minimize this?) What should you do if squirrels or other animals approach the food? (You will need to find a standard way of responding and resuming your observations.)

❧FOLLOWUP:

1.   Tabulate your data for the five observation sessions. What are the possible sources of error in your observations?

2.   Prepare a report of your observations, including the type of food used, the methods you used to make the observations objective, the possible sources of error, and your conclusions. When you have finished, write down at least one question that could be used as the basis for additional study of the food preferences of the birds in your area.

Suggested Bird Foods	
sunflower seeds	unsalted peanuts
niger seed	suet
millet	peanut butter
dry corn	fruit
cracked corn	baked goods

Some Pitfalls to Avoid in a Scientific Study

1. **Observer Effect**: Careful consideration needs to be given to how to reduce any effects of the presence of the observer on the outcome of a study. Viewing animals from a distance through binoculars is one method; building a blind from which the animals can be observed without seeing the observer is another. At the very least, it is important to remain quiet and unobtrusive.

2. **Unintended Effects**: If the observer places the same foods on the same cardboards each time, the birds may be choosing a location, rather than a type of food. A way to avoid this is to **randomly assign** the foods to the cardboards each time; you can put the names of the foods on slips of paper and draw them out one at a time, without looking at them.

3. **Imprecise Terminology**: Terms and procedures need to be described so that other investigators can repeat a study exactly. Defining a visit, or devising a routine on how to proceed if something interrupts an observation, are examples from this study of things that need to be described precisely. The most effective descriptions of terms and procedures are **operational definitions**, that is, descriptions of what a person would need to do (the operations) to recognize a term or repeat a procedure.

Activity #4: Plant Seedlings

In the previous activity, you conducted a field study in which you observed birds in an outdoor setting, collected quantitative data by counting things—the numbers of visits of birds to specific foods—and analyzed the data for patterns from which you could draw some conclusions. Although you tried to make the observations objective by controlling various factors, such as the placement of the foods, there were many factors over which you had no control. The lack of control over some factors is present in any study, but more so when the study is conducted in the field than in a laboratory. In this activity, you will conduct a controlled experiment in your home or school, in which you will be able to manipulate one factor in a methodical manner and control many others. You will also collect data in the form of measurements, rather than counts. Counts are discontinuous, that is, there is nothing between 1 visit and 2 visits (such as 1.34 visits). Measurements, on the other hand, are continuous; you can think of many values between 1 cm and 2 cm (such as 1.34 cm, 1.35 cm, etc.) You can even think of values between 1.34 cm and 1.35 cm (such as 1.344 cm and 1.355 cm).

☞In this activity, you will design and conduct a controlled experiment on the effect of an environmental factor of your choice on the growth of seedlings, measured in length.

✔**MATERIALS:**   10 waterproof cups, such as waxed paper or styrofoam
potting soil
corn, pea, or bean seeds
metric ruler
graph paper (5 squares/inch)
other materials: depending upon the factor you select for your experiment

☉**TIME:**   2 hours for activity, 1 hour for followup

❀ACTIVITY:

1.   Soak the seeds in tap water overnight. The next day, prepare the potting soil according to the directions on the package. This usually involves wetting the soil thoroughly before placing it in the growing containers. Fill the cups 3/4 full with the prepared soil and plant one seed in each cup, following the directions on the seed package as to depth of planting.

2.   When the seedlings begin to push through the soil surface, you should begin your experiment. Select one factor in the environment (such as the type of water, the amount of water, the amount of light) and expose equal numbers of the seedlings to variations of the factor. For example, if you select type of water, you could water some of the seedlings with tap water, some with bottled water, some with acidic water (by adding a small amount of vinegar), some with basic water (by adding a small amount of baking soda), and some with salt water (by adding a small amount of table salt), etc. You could also decide to use various levels of salt water from none (distilled) to very salty. In either case, you would want to keep the amount of water and sunlight constant for all of the seedlings.

3.   On the first day, and on every subsequent day for the next two weeks, measure the length of the seedlings to the nearest millimeter and record the data in a table you prepare.

4.   At the end of the two weeks, construct a bar graph from the data, with the different treatments of the seedlings on the horizontal axis and their length on the vertical axis.

❧FOLLOWUP:

Prepare a report of your experiment, in which you explain the factor you selected and why; how you set up your experiment; what you did to keep factors, other than the one you were manipulating, constant; the results you obtained; and the conclusions you draw from the data. Were there any factors you were not able to control?

POPULATIONS

Activity #5: The Power of Doubling

If you were offered a choice between one million dollars and a penny on the first day of the month, two pennies on the second day, four pennies on the third, and so forth for 30 days, which would you choose? If you calculate this, you will find that the second choice would give you somewhat more than one million dollars! Each day you are doubling a larger number, and although the number of pennies increases slowly at first, it soon reaches over one million dollars. That is the power of doubling, or exponential growth.

Growth is defined as exponential when the increase of a quantity is proportional to the size of the quantity. The quantity may be numbers, as in the numbers of individuals in a population, or some other measure, such as amount of energy consumed. Exponential growth is very slow in the early stages, but quickly accelerates. A frequent measure of exponential growth is **doubling time**, that is, the amount of time required for the quantity to double. The shorter the doubling time, the faster is the rate of growth.

The human population, like all populations of organisms, grows exponentially when unchecked. Although it took 130 years, from 1800 to 1930 for the world population to double, it doubled again by 1975, a mere 45 years. In 1993, the doubling time of the world's population was 42 years. At this rate, the world population of 5.5 billion in 1993 would be expected to reach 11 billion by 2035. Different areas of the world, however, have vastly different doubling times. While the doubling time for developed areas in 1993 was 162 years, that for the less developed areas was 35 years!

☞In this activity, you will demonstrate exponential growth and determine the doubling time and growth rate of a simulated population.

✔MATERIALS: two pennies
approximately 200 dried beans (or dry macaroni)
graph paper (5 squares/inch)
2 8-ounce paper cups
1 larger paper cup
marking pen

☉**TIME:**   1 hour for activity, 1 hour for followup

❀**ACTIVITY:**

1.   Label one small paper cup, "Parents," and the other one, "Offspring." Label the large cup, "Bean Pot." Place 10 beans in the "Parents" cup and the rest in the "Bean Pot." Each bean represents an individual in a population.

2.   Prepare a table consisting of 2 columns and 12 rows. Label the left-hand column "Generation Number" and the right-hand column "Population Size." Number the remaining rows in the left-hand column from 0 to 10. Place a 10 in the right-hand column of the row marked "0" to represent the initial population size.

3.   Toss the two pennies. If both pennies show heads, toss again. If both pennies show tails, one member of the population has died and you should remove a bean from the "Parents" cup and put it into the "Bean Pot." If one head and one tail show, a member of the population has had a child. To simulate the birth, take one bean from the "Parents" cup and one from the "Bean Pot" and place them in the cup marked "Offspring."

4.   Continue tossing until there are no longer any beans in the "Parents" cup. Count the number of beans in the "Offspring" cup and record the number in the data table you have made. The "Offspring" now become the parents, so move all of the beans from the "Offspring" cup into the "Parents" cup.

5.   Repeat steps #3 and #4 until you have completed 10 generations.

6.   Make a graph of your data, with generation number on the horizontal axis and population size on the vertical axis.

✎**FOLLOWUP:**

1.   On average, how should the birth rate of this population compare with the death rate? How do you know this?

2.   From your graph, determine the doubling times for the population at the beginning, the middle, and the end of the graph. Are they all the same? Explain why.

3.   Assume that a generation is equal to 20 years. Use the doubling time from your graph to calculate the growth rate of the population, using the formula below:

annual growth rate (%) = 70/doubling time.

4.   Repeat the activity, beginning with step #2, only this time assume that when a head shows to the right of a tail, the individual decides not to have a child. In that case, place one bean from the "Parents" cup into the "Offspring" cup, but do not add a bean from the "Bean Pot." If, on the other hand, the head shows to the left of the tail, proceed as you did before, taking one bean from the "Parents" cup and one from the "Bean Pot" and placing them in the "Offspring" cup. When you have finished 10 generations, graph your data on the same graph.

5.   Calculate doubling time and growth rate for the second set of data as you did for the first. Compare the two data sets.

6.   Write a paragraph explaining the implications of this activity for the human population.

Population Data for Selected Countries				
Country	Millions (1993)	Growth Rate (%)	% Under 15/Over 65	GNP per capita 1991 (US$$)
Tunisia	8.6	1.9	37/5	1,510
Qatar	0.5	2.1	28/1	15,870
China	1178.5	1.2	28/6	370
Pakistan	122.4	3.1	44/4	400
Uraguay	3.2	0.9	26/12	2,860
Denmark	5.2	0.1	17/16	23,660
United States	258.3	0.8	22/13	22,560

Source: Population Reference Bureau, 1993, *World Population Data Sheet 1993*, Washington, DC: Population Reference Bureau, Inc.

Activity #6: Mark and Capture

Knowing the size of a population of animals is important in making environmental decisions that would affect the population, but estimating the size of wild populations is extremely difficult. In the case of ocean dwellers, such as whales, the task is especially challenging. Estimates of the number of minke whales, for example, have differed by as much as a factor of 10. Deciding whether to allow hunting of minke whales, based on population estimates that are too high, could lead to extinction of the species. On the other hand, basing a decision on an estimate that is too low could unnecessarily ban hunting of minkes by peoples that depend on whales for food. One method for estimating population size, the "line-transect survey," involves observing every animal seen while traveling a straight line. Although traditionally used for counting land animals, the line-transect survey method has recently been applied to whales, providing more reliable data.

Another method often used to estimate population size is the "mark and capture" technique, in which scientists capture some animals from the population, mark them, and release them. At a later time the scientists again capture animals from the same population and observe how many of them are marked. The method assumes that the ratio of the actual population to the sample size is the same as the ratio of the number of marked animals to the number marked in the recapture sample. Knowing three of the four values (sample size, number originally marked, and number marked in the recapture sample), scientists can calculate an estimate of the actual population size.

☞ In this activity, you will use the mark and capture method to estimate the size of a population of pillbugs.

✔ MATERIALS: model paint or opaque nail polish
fine brush
magnifier
meter stick or tape
2 collecting jars or plastic boxes
old boards, brick, or stones (depending on area)

⊘**TIME:**   3 hours for activity, 1/2 hour for followup

❀**ACTIVITY:**

1.   Select an area where the ground is slightly damp and where there is cover, such as stones, leaves, branches, or litter. It is desirable, but not necessary, for the area to be surrounded by a drier area. Use the meter stick or tape to measure the dimensions of the area and estimate its total area in square meters.

2.   Look under the cover and on the surface of the ground for pillbugs. Collect 30-50 and put them in one collection jar. If you have trouble finding pillbugs, you can set bricks or boards on the ground overnight. The next day you should be able to find pillbugs under them.

Pillbug Biology

Pillbugs belong to the phylum **Arthropoda**, which includes insects, arachnids (spiders, mites), millipedes, centipedes, and crustacea. All of these have an outer skeleton, jointed appendages, and segmented bodies. **Crustacea** are a large and varied class of arthropods—its most prominent members being shrimps, crabs, and lobsters—distinguished from other arthropods by having two pairs of antennae. Pillbugs and sowbugs are terrestrial members of the crustacean group known as isopods, most of whose 4,000 species live in water.

Pillbugs and sowbugs have a small head with one pair of large antennae and a second much smaller pair. Behind the head are 7 large (thoracic) segments and 6 smaller (abdominal) segments. Each thoracic segment has a pair of walking legs, and each abdominal segment has a pair of smaller appendages. Pillbugs have an arched back and roll up into a ball when disturbed. Sowbugs are flatter and move away from a disturbance. You can use either for your population study.

The pillbug diet, which is extremely varied, includes fruit, young plants, fungi, decaying matter, and their own feces. The latter habit is related to their need for copper for the oxygen-carrying blood chemical hemocyanin, (as we need iron for our hemoglobin). Pillbug predators include birds, amphibians, mites, spiders, and centipedes. Although pillbugs secrete noxious chemicals from glands along the thorax, their most effective defense is their tendency to seek dark, sheltered places.

3.	Mark the top of each pillbug with a small spot of paint or nail polish. Allow the paint to dry and place the marked pillbugs in the second jar. When all of the pillbugs are marked, release them into the area.

4.	Return to the area every day for a week and capture a sample of pillbugs. Count the number of marked pillbugs in your samples and record the data in your notebook.

5.	Estimate the population size from each day's data by multiplying the number of pillbugs you marked by the number of pillbugs in each sample. Divide the product by the number of marked pillbugs in the sample. After finding the population estimates for each day, find the average estimate by adding up the daily estimates and dividing by the number of samples.

✎FOLLOWUP:

1.	What is the range of the population estimates (highest estimate – lowest estimate)? Is there a trend in the variation in the estimates (for example, do the estimates decrease steadily each day, remain the same, or increase and decrease alternately)? What might explain finding different population estimates?

2.	What assumptions underlie the principle behind the mark and capture method? Are these assumptions valid for your observations?

3.	Using the estimates of population size and area of study, calculate the density of pillbugs (number of pillbugs per square meter).

	Instead of using pillbugs, you can simulate the mark and capture method using marbles.

✔**MATERIALS:**	50 marbles, beads, or other small round objects of the same size
ink or correction fluid that will stick to the marbles
a large paper or plastic container (about 1 gallon), with lid

☉**TIME:**	1 hour for activity, 1/2 hour for followup

✿ACTIVITY

1.   Mark 20 of the marbles with a spot of ink or correction fluid. Allow the markings to dry.

2.   Prepare a table consisting of 3 columns and 11 rows. Label the left-hand column "Trial Number." Label the middle column "Marked Marbles in Sample." Label the right-hand column, "Population Estimate." Number the remaining rows in the left-hand column from 1 to 10.

3.   Place all 50 marbles in the container and cover with the lid. Shake the container to mix the marbles thoroughly. Take the lid off and <u>without looking</u>, withdraw 10 marbles. Count the number of marked marbles. Record the number of marked marbles in the table.

4.   Put the 10 marbles back in the container and replace the lid. Shake the container again and withdraw a second set of 10 marbles. Again, record the number of marked marbles in the sample in the table. Continue in this manner until you have completed 10 trials.

5.   Estimate the population size for each trial by multiplying the number of marbles marked (20) by the number of marbles in each sample (10) and dividing the product by the number of marked marbles for that trial. The result is an estimation of the population size. When you have done this for all 10 trials, find the average by adding the estimates and dividing by 10.

✎FOLLOWUP:

1.   How does the average value compare to the actual population size of 50?

2.   If there is a difference, explain what might cause the difference.

3.   What assumptions underlie the principle behind this method of estimating population size?

4.   What problems might scientists encounter in using this method in the field that you would not have encountered in the simulation?

Activity #7: Estimating Biodiversity

Scientists are concerned that loss of species of living organisms as a result of human activities is endangering the stability of the biosphere. The rate of extinction accelerated during the past century, and, today, many more species are in danger of or threatened with extinction. It has been difficult to determine the exact scope of the problem, however, because scientists do not know how many species there are at present, much less how many existed 100 years ago.

Almost 1,400,000 species of living organisms have been described, but scientists believe the actual number to be much greater. Recent studies in the Peruvian rain forest have uncovered an unanticipated degree of diversity among insects. Some scientists estimate that the total number of species on Earth may be as high as 30 million, if all of the rain forests of the world have a similar diversity. The majority of scientists, however, estimate the number of species as between 5 and 10 million.

A common technique for estimating the total number of species in an area is to count the number of species in successive samples within the area. When a point is reached at which additional samples do not yield additional species, it is assumed that all, or almost all, of the species have been uncovered.

☞In this activity, you will estimate the total number of species in an area near your home or school.

✔**MATERIALS:**   4 meter sticks[1], or 4 sticks 1 m long, or 1 hula hoop
tape (if using sticks)
string
2 small sticks
magnifying lens
graph paper

⊘**TIME:**   2 hours for activity, 1 hour for followup

---

[1]Inexpensive meter sticks are usually available at lumber yards or sewing stores.

→ You can use the same methods for Activity #21: Succession.

❀ACTIVITY:

1.   If you are using sticks, construct a square with sides of 1 meter by taping the sticks together at right angles to one another. Instead of a square, you can use a hula hoop. You will also need to select the area in which you are going to make a census of the number of species. It should be at least 10 meters long and 2 meters wide.

2.   Prepare a table consisting of 3 columns and 11 rows. Place labels on the top row as follows: left-hand column "Sample Number," the second column "Number of New Species," the third column "Cumulative Number of Species."

3.   Tie one end of a string to a small stick at one end of your study area and the other end to another stick at least 10 meters away. Place your square or circle flat on the ground at the first stick, with the area of the square or circle lying in the direction of the second stick. Carefully identify every species of plant and animal within the area of your square or circle. You may have to look closely at ground level with the magnifying lens to do this. If you are not able to identify species by scientific name, you can make up your own names. Collect samples of each species or make accurate drawings of characteristic parts so you can compare them with organisms in subsequent samples. Record the total number of species for sample #1 in the second and third columns.

4.   Flip your square or circle over so that it lies along the string and further in the direction of the second stick. The area of this second sample should not overlap that of the first sample. Again identify all of the plants and animals within the area of your square or circle. Record the number of new species in the second column and the cumulative number of species (number of species from sample #1 plus number of new species from sample #2) in the third column. Continue in this manner until you have reached the second stick or until you find no new species for several consecutive samples. If you are still finding new species when you reach the second stick, set up another line that does not overlap your first line and continue taking samples.

5.   Plot your results on a graph with the sample numbers on the horizontal axis and the cumulative number of species (the number of new species at that sample plus the number of species identified in previous

samples) on the vertical axis. Estimate the total number of species for your study.

FOLLOWUP:

1.   What are the possible sources of error for your estimate? Assuming your estimate is very accurate for the area you studied, how accurate would it be for adjacent areas? for different habitats?

2.   What are the possible sources of error in the estimate of the number of species of living organisms?

3.   You can calculate a crude index of species diversity by dividing the total number of species found by the total area. Calculate total area (number of samples x area of each sample) and then the species diversity. Compare your index of diversity with those found by other students. Do any patterns emerge when you compare diversity with type of habitat?

Number of Living Species Known	
Group	Number of Species
viruses	1,000
bacteria and similar organisms	4,800
fungi	69,000
algae	29,000
protozoa	30,800
higher plants	248,400
insects	751,000
other animals	281,000
total	1,413,000 [sic]

Source: Reprinted by permission of the publishers from *The Diversity of Life* by Edward O. Wilson, Cambridge, Mass.: The Belknap Press of Harvard University Press, Copyright © 1992 by the President and Fellows of Harvard College.

Activity #8: Making an Age Pyramid

Population growth is affected by age structure—the number of people in different age groups—as well as by the numbers of births and deaths. Age structure is usually illustrated by an age pyramid, a graph in which horizontal bars represent the percentage of the population in each age group. Males are shown on the left and females on the right. The ages (or in some cases, the years of birth) for each bar are listed along the vertical axis of the graph, usually in five-year intervals. Each age group is called a cohort. The longer a bar is, the greater the proportion of people in that age group.

Age pyramids are useful for tracing the history of a country's population and for projecting future population trends. A very long bar for the age group of 40-45 years olds would indicate that a large number of babies had been born between 1940 and 1945. An age pyramid with more long bars for the younger age groups would indicate a growing population; when these large numbers of young people begin to reproduce, they will add even more children to the population than did the older age groups.

☞ In this activity you and your classmates will collect data on your families and pool the data to produce an age pyramid diagram. This will help you to interpret age pyramids and understand the relationship between age structure and population growth.

✔MATERIALS: graph paper, 5 squares/inch
                 class data

☺TIME: 2 hours for activity, 2 hours for followup

❀ACTIVITY:

1.   Collect information to complete the "Family Population Form" (see page 35). Find out the birth date and sex of each member of your family, beginning with your grandparents. Include all of the brothers and sisters of your parents and all of the people in your generation, i.e., your brothers and sisters and cousins.

2.   Pool your data with that of your classmates. Construct an age pyramid diagram for the class data using graph paper with 5 squares to the inch. You will need to decide how many people are to be represented by one square. The example for Iran below should help you.

Iran, 1986

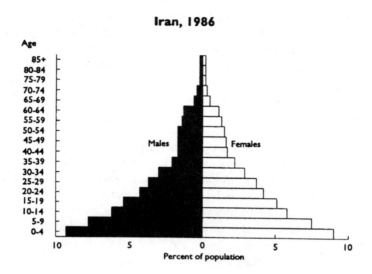

Source: Abdel Omram & Farzaneh Roudi. The Middle East Population Puzzle, *Population Bulletin*, vol. 48, no. 1 (Washington, DC: Population Reference Bureau, 1993).

✎FOLLOWUP:

1.   What is the percentage of people under 20? over 60? Is the population a young, growing one; an older, declining one; or a stable one?

2.   Is there evidence in the diagram of the baby boom that followed World War II (1946-1964)? If so, is there evidence of the effect of this baby boom in more recent years?

3.   From the class data, determine the average number of children per couple for each generation and compare the averages to the replacement level of 2.1 children per couple. Describe any changes that have occurred in family size over the generations.

4.   How does the pyramid for the class compare to that for the United States? If the United States is now at, or slightly below, the replacement level, why is the population of the country still growing?

FAMILY POPULATION FORM

Date of Birth	Number of Females	Number of Males
1995-1999		
1990-1994		
1985-1989		
1980-1984		
1975-1979		
1970-1974		
1965-1969		
1960-1964		
1955-1959		
1950-1954		
1945-1949		
1940-1944		
1935-1939		
1930-1934		
1925-1929		
1920-1924		
1915-1919		
1910-1914		
1905-1909		
1900-1904		

HABITATS & COMMUNITIES

Activity #9: A Grass-Roots Community

Communities are composed of the plants and animals living in a particular habitat. Forests, meadows, tide pools, and ponds are communities. Communities exist in urban areas, but they are often overlooked. The organisms in a community are adapted to the physical environment and to the other organisms in the habitat. Certain types of organisms are found in every community: those that are able to produce their own food (producers), and those that depend upon other organisms for food (consumers). Those consumers that feed on dead or decaying plants or animals are called decomposers. They break down once-living organisms to their component materials and return them to the environment, where they can be used by a new generation of living organisms.

☞In this activity, you will study and compare the microclimates and living organisms in two microcommunities.

✔MATERIALS: string
                wooden or plastic stakes
                meter stick
                thermometer
                small digging tool
                graph paper
                field guides (optional)

☺TIME: 2 hours for activity, 1 hour for followup

❀ACTIVITY:

1.   Select an area for study where two habitats are adjacent to one another, such as a lawn area at the edge of a woods. Follow the procedure described in the next steps for each habitat.

2.   Mark off an area 1 meter square with stakes and string. Make a scale map of the area on graph paper. Include special features, such as rocks and trees.

3.   Make notes on the general characteristics of the area, such as type of soil, amount of shade, proximity to sidewalks or roads, etc.

4.   Measure the temperature at each of the following points: 1 meter above ground, at ground level, 2.5 cm below ground level, and 7.5 cm below ground level.

5.   Dig down 10-20 cm below ground level to expose the soil layers. Draw a diagram of the different layers, indicating the color, moisture level, and texture at each level.

6.   Classify the plants in the area according to type: lichen, fungus, moss, fern, tree seedling, grass, small broad-leaved plant (stems not woody), shrub (woody plant with several to many stems starting at or near ground level), tree (woody plant with one main trunk). The most numerous plant type is the dominant one. List the types of plants you find.

7.   Lay the meter stick along one side of the study area and look along the inside edge for its entire length for animal life. Look above, at, and below the surface. Turn over rocks, leaves, or other debris. When you have finished, move the meter stick 10 cm toward the inside of the study area and observe the animals found along this new line. Continue in this manner until you reach the opposite side of the study area. Identify and list the organisms.

✎FOLLOWUP:

1.   What differences in animal and plant life did you find in different parts of each area? How do you think these differences are related to variations in temperature and moisture in the area?

2.   Prepare a brief report comparing the communities found in the two habitats. Illustrate the report with the maps, tables, and any drawings you have made.

Activity #10: Wildlife in Your Area

Traditional definitions of wildlife have focused on large animals desired by hunters. Recently, the definition has been broadened to include endangered species and animals important for tourism. However, any species not cultivated by humans can be considered "wildlife." Undesirable wildlife, such as rats, and those that come into conflict with humans, such as wolves preying on domestic cattle, are often labelled as "pests," much as undesirable plants are called "weeds."

Using a broad definition, wildlife can be found in almost every environment, even those most strongly affected by human activities, the cities. Wild animals can be found wherever they can meet their survival needs: food, shelter, water, mates, and places to rear young. The purpose of this activity is to increase your awareness of the living resources around you, and of the relations between humans and living resources.

☞In this activity, you will conduct a survey of your neighborhood for signs of wildlife.

 You may wish to do this activity in conjunction with the next one.

✔**MATERIALS:** camera (optional)

☾**TIME:** 3 hours for activity, 1 hour for followup

❀**ACTIVITY:**

1. Conduct a preliminary survey of your campus or neighborhood to identify places where animals might find what they need to survive. Select two areas of about the same size, one relatively undisturbed and one relatively disturbed by people. (Most campuses have some heavily vegetated and/or neglected, out-of-the-way areas.) Prepare a brief, written description of the two sites you have selected, noting particularly evidences of human impact.

2. Make at least three one-hour observations at each site, one in the very early morning, one in the middle of the day, and one around dusk. Try

to make your observations on two consecutive days. Make a record of the kinds of animals you see or hear and the numbers of each kind. Indicate which ones seem to live at the site and which appear to be visiting. Determine, as best you can, how each of the former would be able to meet its survival needs at that site. For each of the latter, determine, as best you can, which survival need (or needs) the animal is finding at the site.

3.   If you do not actually see any animals, search the area carefully for sources of food, water, and shelter, and from these observations, predict the kinds of animals that may be living in the area. Some of the signs you might look for are:

food: grass, leaves, twigs, berries and other fruits, seeds, nuts, grains, flowers, small mammals and birds, insects, decaying plants and animals, garbage

water: ponds, puddles, streams, bird baths

nesting sites: cliffs or high buildings, cavities in trees or buildings, trees, shrubs, depressions or holes in ground, ponds, streams

✎FOLLOWUP:

1.   Based on your survey, what types of wildlife would you expect to find?

2.   Compare your observations of the two sites and make as many inferences as you can about the reasons for differences between them.

3.   Share your findings with others from your class. Are there any findings common to many groups? Do these lead to further inferences?

4.   Make a list of ways in which either site could be improved as a habitat for wildlife. What advantages or disadvantages would implementing these recommendations have for the people in the area?

Activity #11: Improving a Wildlife Habitat

Although humans have had a major impact on many wildlife habitats, they are also beginning to appreciate the need to restore disturbed areas and to learn how they can live more in harmony with other life forms. Even urban areas provide the basic survival needs—air, water, food, and a place to live—for a variety of wildlife. Wildlife benefits people in many ways. Plants moderate some nonliving factors in the environment, such as water supply, climate, wind, evaporation, and air quality; they provide many products for human use as well. Animals, too, provide people with food, and also have many other uses. But people do not see wildlife only in utilitarian terms. People enjoy having living organisms around them. Pets and gardens are ways in which urban people compensate for being more isolated from wildlife than their ancestors were. Parks and zoos are popular also because they bring people in contact with wildlife. Often, however, people do not see the wildlife in their midst, or the possibilities for bringing more wildlife into their own neighborhoods.

☞In this activity, you will select an area and propose ways to improve it as a habitat for wildlife.

✔**MATERIALS:**   camera, videocamera, cassette recorder, or drawing supplies (optional)

☺**TIME:**   2 hours for activity, 1 hour for followup

❀**ACTIVITY:**

1.   Identify a site in your community that can be restored and improved as a wildlife habitat, such as an abandoned site, vacant lot, drainage ditch, farm pond, grove of trees, or roof top.

2.   Survey the site and complete a "Wildlife and Habitat Inventory" (see page 45) for the site. You may wish to document your survey with maps, drawings, or photographs.

✎FOLLOWUP:

1.   Evaluate the suitability of the site as a wildlife habitat.

2.   Find out about local laws, ordinances, and environmental regulations that control land use in your area.

3.   Develop a proposal for improving the site as a wildlife habitat. Refer to the list of "Plants Recommended for Attracting Wildlife." You can also obtain information from a county agriculture extension office or local garden store. Include in your proposal:

   a.   recommended plants to be added to the site
   b.   addition of water sources, such as a small pond or bird bath
   c.   construction of shelter and living space, such as rock or brush piles, bird houses and feeders
   d.   proposed facilities for people to enjoy the site, such as benches, refuse containers
   e.   suggestions for continuing maintenance and management.

You may want to select a site that has already been improved for wildlife and report on what was done to restore the area and how it is managed. In recent years, pocket parks and community vegetable gardens have become popular ways of using neglected areas in cities. Some real estate developments have included green areas and federal regulations prohibit destruction of wetlands. In some cases, utility companies are required to protect wildlife along power lines, reservoirs, and aqueducts. Oil fields, military bases, and waterfront areas often become sanctuaries for wildlife. Another alternative is for you to adopt a small area to improve or restore, or for your class to take on a large area as a class project.

Wildlife and Habitat Inventory

Plants	Types	Number*
	trees	
	shrubs	
	vines	
	grass	
	flowers	
Food	berries	
	seeds	
	nectar	
	worms	
	insects	
	other	

Water	Type	Yes/No
	pond	
	pool	
	stream	
	bird bath	
	other	
Other Cover	dead trees	
	tall grass	
	gravel	
	rock pile	
	brush pile	
	other	

*None, few, or many?

Estimate the following: % of area covered by trees and shrubs_____

                              % of area covered by buildings and pavement _____

                              % of area covered by grass   _____

List the animals you have seen.

Adapted from "Wildlife and Habitat Inventory," National Institute for Urban Wildlife, 1991. *Developed Lands: Restoring and Managing Wildlife Habitats* (Teacher's Pac series in Environmental Education). National Institute for Urban Wildlife, P.O. Box 3015, Shepherdstown, WV 25443. Copyright © 1991, National Institute for Urban Wildlife.

Plants Recommended for Attracting Wildlife

1. Large trees

beech	EMW	pine (Eastern white, Scotch)	EM
fir, Douglas	W	pine (Western white, ponder-	
eastern hemlock	E	osa, lodgepole)	W
red maple	EM	sassafras	EM
white oak	EM	sweetgum	EM
pin oak	M	American sycamore	EM
California black oak, Oregon		tulip-poplar	EM
white oak	W		

2. Small trees

black cherry	EMW	mesquite	W
crabapple	EMW	mulberry (red, white)	EMW
flowering dogwood	EMW	redbud	EM
hackberry	EMW	Eastern red cedar	EM
Washington hawthorn	EMW	Western red cedar	W
American holly	EM	serviceberry	EMW

3. Large shrubs

high-bush blueberry	EMW	bush honeysuckle	EMW
common buttonbush	EM	Oregon grape	W
cascara	W	manzanita	W
black chokeberry	EMW	multiflora rose	EMW
coralberry (common		autumn olive	EMW
winterberry)	EMW	Russian olive	EMW
dogwood (silky, gray)	EMW	Amur privet	EM
Elaeagnus cherry	EMW	sumac	EMW
elderberry	EMW		
firethorn (pyracantha)	EMW		

4. Small shrubs

Japanese barberry	E	bicolor lespedeza	EMW
bayberry	E	pokeberry	E
American beautyberry	EM	common snowberry	EMW
buffaloberry	W	viburnum (arrowwood,	
cotoneaster	EMW	nannyberry, American	
Japanese holly	EMW	cranberry)	EMW

Key: E = East; M = Midwest; W = West

Adapted from "Wildlife and Habitat Inventory," National Institute for Urban Wildlife, 1991. *Developed Lands: Restoring and Managing Wildlife Habitats* (Teacher's Pac series in Environmental Education). National Institute for Urban Wildlife, P.O. Box 3015, Shepherdstown, WV 2544. Copyright © 1991, National Institute for Urban Wildlife.

5. Vines and ground covers (E, M, W)

bearberry	ground juniper
bittersweet	Virginia
common	creeper
trumpet-	wild grape
creeper	

6. Flowers (E, M, W)

aster	jewelweed
columbine	lily
coral root	marigold
cosmos	petunia
crysanthemum	sunflower
forget-me-not	zinnia
hollyhock	

Urban Naturalists

The letter below describes an example of what can be learned by observing wildlife in a city. It was written to the editor of The New York Times, and appeared in the March 8, 1994 issue (p. A-20). The writers , Cindy Moses and Lucinda Randolph, are a former student in one of the New York City 4-H Clubs, who now recruits for the program, and the director of the program, respectively. They have graciously granted permission for their letter to be reprinted here.

To the Editor:

We New York City 4-Hers are not as surprised as the scientists that male bats lactate (news article, Feb. 24). As part of our pigeon project in the Bedford-Stuyvesant section of Brooklyn, we learned that male pigeons lactate and also stay in the nest during the day while the female hunts for food. They also mate for life!

New York City 4-H Clubs are part of Cornell Cooperative Extension. The pigeon project helped us find beauty in our own neighborhood and taught us respect for city wildlife.

Cindy Moses, Lucinda Randolph
Brooklyn, Feb. 25, 1994

Activity #12: Community Relations

Patterns of behavior that promote survival of offspring contribute to the evolutionary success of a species. Natural selection results in members of a species developing adaptations, including behavior patterns, to their environment. Thus, the behavior of organisms provides clues as to how they interact with their environment. Using the study of behavior to link evolution with ecology is known as behavioral ecology. Observing food gathering and relations with members of the same and other species—including mating, care of young, prey-predator relations, territorial behavior, and social behavior—are ways of learning about how organisms relate to their ecological community. The role that a species plays in its community is called its ecological niche. A habitat is where a species lives; its niche can be thought of as how it makes its living.

☞In this activity, you will use observations of a species of animal to infer its role in its ecological community.

✔MATERIALS:　　watch
binoculars (optional)
camera or videocamera (optional)

☺TIME:　　2 hours (minimum) for activity, 1 hour for followup

❀ACTIVITY:

1.　　Select an animal or bird found in your area as a subject. You may need to do some preliminary observations to decide which subject to study and the best area for doing so. Squirrels are common throughout the country and, unlike most mammals, are active during daylight. Rabbits, chipmunks, house sparrows, robins, starlings, blue jays, cardinals, and pigeons also make good study subjects. You will need to sit or walk quietly and avoid sudden movements. You should plan on at least 2 hours of observation; however, you may want to spread this out throughout the day or over several days. The more observations you can make, the more you will learn about your subject. If it is possible for you to work with a group of students, you can coordinate your observations and pool your results.

2.   Make a sketch of the study area, including prominent features, such as large trees, buildings, lawn areas, etc. At each observation session, note the weather conditions, such as temperature, wind, amount of sunlight, and time of day.

3.   Look for patterns of activity, such as which parts of the study area are frequented by your subject, what it feeds on, how it relates to other members of its species and to other species. You will be better able to infer patterns if you tabulate repeated behaviors. For example, you might make note of the number of times a squirrel sits up on its back legs and looks around. You will find it useful also to use abbreviations, numbers, and symbols for individuals and behaviors. When you can identify different categories of behaviors, such as eating, climbing, etc., keep a record of the time the animal spends in each category. At the time, you may not be able to detect a pattern, but by reviewing your accumulated data, you may find that one exists.

✎FOLLOWUP:

1.   Based on your observations, estimate the size of the population of the subject species in that area. What sources of error are there in your estimate? How could you improve your estimate?

2.   Summarize the animal's behavior by calculating the percentage of time spent in the categories of behavior you observed. Relate your results to the animal's habitat and niche. If you have made observations at different times of day, compare behavior patterns at these different times. Compare the time budget for the organisms you observed with those observed by your classmates for the same or different species. Try to account for differences in the ways different organisms spend their time.

3.   Write a report summarizing your observations and describing the habitat and as much as you can about the role of the species you studied in its community. Include map, data tables, graphs, and drawings.

ECOSYSTEMS

Activity #13: Creating a Mini-Ecosystem

Creating a large, self-contained ecosystem that will sustain itself indefinitely, as the designers of Biosphere 2 (see box on page 54) attempted to do, is a scientific and technological challenge still in the experimental stages. No one knows for certain how to do it or whether, in fact, it is possible to maintain an artificial ecosystem for long periods of time. In the process of trying to find out, however, scientists are learning more about natural ecosystems.

The three most basic elements in an ecosystem are an outside source of energy, a form of life that can capture the energy, and a supply of matter that can be recycled within the system. For most of the ecosystems on Earth, sunlight is the energy source. Green plants or algae capture the sun's energy and convert it into living matter. Animals depend on the ability of producers to make living matter. Microorganisms recycle the living material into nonliving material, which can be reused by the producers.

☞ In this activity, you will create a small closed ecosystem and observe it for several weeks.

✔**MATERIALS:**   1 glass quart jar, with screw-type lid
pond water or seawater
sediment or other substrate material from pond or
tide pool
plants and animals from pond or tide pool

☺**TIME:**   1 hour to set up ecosystem, plus a few minutes observation time each day for two months; 1 hour for followup

❀**ACTIVITY:**

1.   Fill the glass jar with water from a pond or tide pool. Add a few small fish, crustacea, snails, plants, and sediment or pebbles. Recycling in ecosystems is carried out largely by microscopic organisms. Even though you won't be able to see them, you will be adding them to your ecosystem with the water and sediments.

2.   When you are finished, seal the jar tightly and place it in **indirect** sunlight. Observe the jar every day for at least two months, making notes on your observations in a log. Do not be discouraged if nothing seems to occur for several weeks, but be sure to note even the most subtle changes. If you have the facilities to do so, set up several jars with different amounts of living matter and different combinations of organisms. Make careful note of the initial differences in types and abundances of organisms so that you can try to account for any differences in results.

✎FOLLOWUP:

At the end of the two months, write a report of two to three pages on your observations. Attach your observation log.

Biosphere 2

Biosphere 2 is the most ambitious attempt yet to build a large, self-sustaining ecosystem that will support humans. It began with experiments on mini-ecosystems by Clair Folsome at the University of Hawaii, who sealed sea water and marine organisms in 1.5 liter glass containers. (Some of those he sealed in 1968 are still going.) Scientists study self-sustaining ecosystems for what they can learn about the biosphere of Earth and the design of space ships that could support humans on long voyages.

Biosphere 2, which occupies 1.28 hectares in the Arizona desert, includes different biomes (desert, ocean, marsh, rainforest, savannah) and over 3,000 species of living organisms (not counting microorganisms in the soil). All wastes and inedible materials were to be recycled.

In 1991, after seven years of planning and $150 million, eight men and women began a two-year stay in Biosphere 2. Although they were able to raise 80% of their food, they had to spend most of their time doing it, and still suffered from hunger and weight loss. Too much carbon dioxide and too little oxygen gave them headaches and shortness of breath. The populations of algae and cockroaches exploded, while those of pollinating bees died off.

In addition to the technical and ecological problems, Biosphere 2 has been attacked for being more theatrical than scientific, and for being secretive when the spirit of science is an open exchange of ideas. Conflict among participants in the project has further compromised the Biosphere's reputation. In 1994, charges of sabotage were brought against two former inhabitants of the Biosphere, who claimed to be protesting mismanagment of the project. So far, Biosphere 2 has proven how difficult the challenge of building self-sustaining ecosystems is. Whether it will lead to a greater understanding of ecosystems is still open to question.

Creating a Micro-Ecosystem

You can observe microscopic producers, consumers, and decomposers in pond water, track changes in the ecosystem over time, and determine how long it can survive with no inputs other than sunlight.

✔MATERIALS: slides and cover slips
dropper
glass or plastic container for pond water (small T-flasks, used in tissue culture, make good containers)
guide to microscopic freshwater organisms

☉TIME: 1 hour to set up ecosystem, plus 1 hour a week to make censuses; 1 hour for followup

✿ACTIVITY:

1.   Collect pond water, including some algae and other pond plants, in a container and put it where it can get light. If you have a T-flask, fill it with pond water and keep the large container as a stock culture.

2.   If you are using a T-flask, put it on the stage of the microscope and focus on the pond water at 100x magnification. Identify as many types of microorganisms as you can, or draw diagrams to represent them. If you do not have a T-flask, put a drop of pond water on a slide, cover the drop, and identify organisms. You should move the microscope to different areas of the T-flask or take drops from different parts of the container of pond water to be sure you have seen a representative sample of the organisms.

3.   After you are familiar with the organisms, do a census of them by counting the number of each type of organism you see in one field of view. Move to another area of the T-flask, or to another drop from the container, and repeat this procedure until you have 10 samples.

4.   Repeat the census every 3-4 days for a period of several weeks.

 You can use the changes in the micro-ecosystem in pond water to study succession, as described in Activity #21.

⬟FOLLOWUP:

At the end of several weeks, write a report of your observations, including graphs of the census results. Develop one or more hypotheses to explain the changes you observed.

Activity #14: Food Web Interactions

Flathead Lake in Montana illustrates what can happen in an ecosystem when a new species is introduced. Opossum shrimp were not found in the lake until 1981, when they moved into it from rivers feeding the lake. They had been introduced into the rivers between 1968 and 1975 as food for kokanee salmon. The shrimp feed on two groups of small floating (planktonic) crustacea—copepods and cladocera—in surface waters at night, but spend the days at depths of 30 m or more. Kokanee feed primarily in the day at depths of less than 30 m. Consequently, kokanee did not feed on the shrimp, although deeper feeding fish, such as lake trout and whitefish did. The figure below illustrates the organisms primarily affected by the introduction of the opossum shrimp.

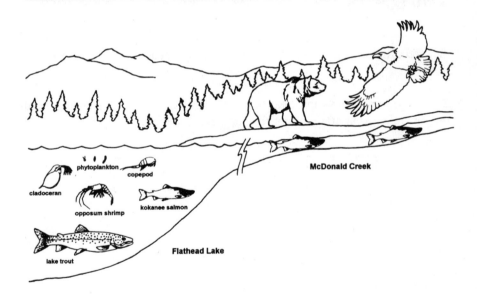

Organisms of the Flathead Lake Ecosystem

Source: C. N. Spencer, R. McClelland, & J. A. Stanford, 1991 (Jan.), Shrimp Stocking, Salmon Collapse, and Eagle Displacement, *Bioscience*, *41*(1): 14-21. Copyright © 1991 by the American Institute of Biological Sciences. Reprinted by permission.

☞In this activity, you will analyze the food web in Flathead Lake and predict the effects of the introduction of opossum shrimp.

☺TIME: 1/2 hour for activity, 1 hour for followup

❀ACTIVITY:

1.   Study the diagram, "Food Web for Flathead Lake," which represents the major organisms found in the lake prior to 1981. The arrows point from prey to predator. Make a table with two columns. In the left-hand column, list all of the organisms shown in the food web.

2.   Using dotted lines, add the opossum shrimp to the web, showing what it eats and what eats it.

✎FOLLOWUP:

1.   In the right-hand column of the table you made in step #1, predict the effect that the opossum shrimp would have on each organism in the food web.

2.   If runoff into the lake began to increase, bringing more nutrients into the lake, what effect might this have on the system?

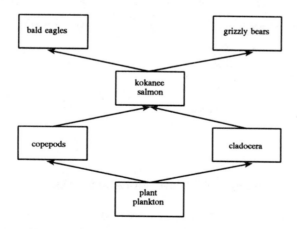

Food Web for Flathead Lake

Activity #15: Owl Pellets

Owls are predators that feed on small mammals, birds, and caterpillars. Because birds do not have teeth, owls use their sharp beaks and talons to tear the prey into large pieces, or swallow it whole. The bones, fur, and feathers are not digestible, so once or twice a day they regurgitate these in the form of a compact pellet. The shape and size of the pellets can be used to identify the species of owl and their prey can be identified by examining the bones in the pellets. Other types of birds, such as hawks, eagles, kites, harriers, falcons, crows, and robins also produce pellets.

Pellets are not the same as feces, which consist of undigested material that has passed through the digestive tract and been eliminated through the anus. The pellet you use in this activity will have been fumigated to kill any living insects that may be in the pellet. The pellet is clean and the only possible safety problem is that some people are sensitive to the fur and dust in the pellet. If this applies to you, you can purchase an inexpensive dust mask in a grocery or hardware store.

☞In this activity, you will investigate the eating habits of owls by examining their pellets and estimating the number of prey they eat in a year.

✔**MATERIALS:**   1 owl pellet (from your instructor, or see
                              "Sources of Information and Materials"
                              in the back of book)
                       1 small glass or glass jar
                       metric ruler
                       metal or bamboo skewers (or other sharp tools)
                       magnifying lens
                       6 5" x 7" index cards
                       glue stick
                       fine mesh strainer or small piece of mesh cloth
                       several paper towels
                       chlorine bleach (optional)
                       graph paper (5 squares/inch)

☾**TIME:**   2 hours for activity, 1 hour for followup

❦ACTIVITY:

1.   Note the color, texture, shape, and size of the pellet. Make a sketch of the pellet and record its length and width.

2.   Using the skewers, <u>carefully</u> tease the pellet apart. Pick out the bones and separate them from the hair. Put the bones in the jar with some water. Any hair that you could not separate from the bones should float to the top. (If you wish the bones to turn out white, you can add one or two drops of bleach to the water. **Caution:** Too much bleach will dissolve the small bones.)

3.   Pour off the water from the top to remove most of the hair. Add more water to the jar and continue to pour off until there is no hair left. Strain the water to separate the bones, and place them on a paper towel to dry.

4.   Use the "Owl Pellet—Bone ID Sheet" (see next page) to identify the bones of the skull, teeth, legs, ribs, and feet. Separate the bones according to size and identify the animals from which they came. Paste the bones from each type of animal on an index card.

❧FOLLOWUP:

1.   Estimate the number of animals of each type in the pellet. The best basis for this estimate is the number of skulls.

2.   Owls usually cough up two pellets every day. Assuming this is true for your owl, how many prey did it eat in a day? How many would it take to feed the owl for a year?

3.   Use the following information on approximate weight of owl prey to estimate the biomass needed to support the owl for one year: rodents (mice) - 25 g; shrews - 7 g; moles - 100 g; birds - 20 g.

4.   Estimate the biomass required to support the owl's prey for one year. (Assume that the prey are 10% efficient in storing biomass.)

5.   Draw a food pyramid **to scale** showing amount of biomass at each level.

Owl Pellet - Bone ID Sheet

	Skulls & Jaws	Scapula	Front Legs	Ribs	Vertebrae	Pelvis	Hind Legs	Feet
RODENTS								
SHREWS								
MOLES								
BIRDS								

Source: Courtesy of Genesis, Inc., P. O. Box 2242, Mount Vernon, WA 98273. Copyright © 1994 *Genesis Inc.*

Activity #16: Carrying Capacity

The productivity of an ecosystem limits its carrying capacity, that is, the mass of living organisms that the ecosystem can support. The carrying capacity of the Earth usually refers to its ability to support human life, because it is the human population that is currently undergoing explosive exponential growth. But the concept of carrying capacity can be applied to any life form and to any part of the biosphere, such as the number of deer that can be supported by an oak forest. Acorns, produced by oak trees, are a favorite food for deer, as well as for squirrels, jays, quail, crows, woodpeckers, raccoons, rabbits, and foxes.

In areas with mild winters, acorns may be available for 8 months of the year and constitute about 75% of the diet of deer. Acorns are higher in fat and easily-digested carbohydrates than other food sources, such as leaves, twigs, small green plants, and fungi. In areas with hard winters, reproductive success of deer decreases with greater snow cover, when acorns may harder to find. Deer have reduced birth weights and lower survival of fawns when acorns are less available. In areas with mild winters, such as the southeastern United States, deer appear to be better able to survive years of low acorn production by shifting to other foods.

☞In this activity, you will create a model of an oak forest and estimate the number of deer that can be supported by the forest.

✔MATERIALS: graph paper

☺TIME: 2 hours for activity, 1 hour for followup

✿ACTIVITY:

1.   Use the data in the table, "Acorn Yield Per Year" (see next page), to make a graph of acorn yield in kilograms (vertical axis) versus diameter at breast height (centimeters) for the five species of oak. It will be easier to read the graph if you use a different color of pen or pencil for each species.

2.   Using the information in "Oak Species in Virginia" (see next page), design a forest of oak trees that will yield a maximum supply of acorns. Assume a density of 25 oaks per hectare and select the species and diameter

of each tree. Use the data on acorn yield to calculate the acorn potential for the forest.

Oak Species in Virginia

Common Name	Scientific Name	Habitat
white oak	*Quercus alba*	dry or moist woods
post oak	*Quercus stellata*	dry soils
blackjack	*Quercus marilandica*	dry, barren soils
Spanish oak	*Quercus falcata*	woods
water oak	*Quercus nigra*	coastal plain

Acorn Yield Per Year (kilograms)

Diameter (cm)	Oak Species				
	white oak	post oak	black-jack	Spanish oak	water oak
10	------	0.3	------	------	------
15	------	0.6	------	------	------
20	0.2	1.0	------	0.5	0.7
25	1.2	1.3	0.8	1.4	1.8
30	2.2	1.6	1.5	2.3	3.1
35	3.2	1.9	2.2	3.2	4.2
40	4.2	2.3	3.0	4.1	5.4
45	5.2	2.6	3.7	5.0	6.6
50	6.2	3.0	4.6	5.9	7.8
55	7.2	3.3	5.2	6.7	9.0
60	8.2	3.6	5.9	7.6	10.1
65	9.2	4.0	6.7	8.5	11.3

Source: P. D. Goodrum, V. H. Reid, & C. E Boyd, 1971 (July), Acorn Yields, Characteristics, and Management Criteria of Oaks for Wildlife, *Journal of Wildlife Management,* 35(3): 520-532. Reprinted by permission.

3.   Assuming that the average deer requires 3 kilograms of food a day and that 75% of the diet is acorns, calculate how many deer each hectare of your forest could support for a year.

4.   Scientists estimate that about 15% of the acorn yield is eaten by birds and others that feed in the trees; only 85% reaches the ground. Adjust your calculations to take this factor into account.

✎FOLLOWUP:

1.   What is the relationship between diameter of oaks at breast height and acorn production? If information were available for trees greater than 65 centimeters in diameter, what would you predict for their acorn production? (Blackjack and post oaks rarely grow over 70 centimeters in diameter, and the others rarely over 90 centimeters.) Based on the information about the species, can you offer a hypothesis about why some species produce greater acorn yields than others?

2.   What is the carrying capacity of the forest for deer? How does it compare with the carrying capacity found by other people doing this activity? After comparing your results with theirs, would you modify your model in any way? If so, how?

3.   How would the presence of other animals that eat acorns from the ground affect the number of deer the forest can support?

4.   Squirrels are more dependent upon acorns as a food source than are deer; that is, they have fewer alternative food supplies. How might a high density of deer in an area affect the population of squirrels?

5.   Although squirrels can usually find the acorns they have buried, some escape. Deer eat acorns directly from the ground or trees, without burying them. How might succession in a forest that had deer, but no squirrels, differ from one that had squirrels, but no deer?

CYCLES & SUCCESSION

Activity #17: Plant Productivity

A self-sustaining ecosystem requires that matter be recycled. The cycles of the four most abundant elements in living matter—carbon, hydrogen, oxygen, and nitrogen—are especially important. The biological carbon cycle depends upon photosynthesis and respiration. In photosynthesis, producers—green plants and algae—combine carbon dioxide with water to produce sugar and oxygen. The producers give off the oxygen to the atmosphere or water in which they live and use the sugar as an energy source to power all of their life activities. To do this, they carry out respiration—the biochemical process in which oxygen is used to break down sugar, releasing energy, water, and carbon dioxide. Consumers also depend upon respiration for the energy they need to carry out their life activities. The complementary processes of photosynthesis and respiration cycle carbon through the living world.

Photosynthesis is basically an energy-storing process in which energy of sunlight is converted to chemical energy. The rate at which the sun's energy is converted into chemical energy by producers is called the **primary productivity** of an ecosystem. Because some of the chemical energy produced in photosynthesis is used by the producers themselves in their own respiration, only the **net productivity** is available for consumers. Consumers cannot make their own food; therefore, the net productivity of an ecosystem is entirely the work of the producers. Consumers as well as producers carry out respiration, however, so the total respiration for an ecosystem includes that of producers and consumers.

☞In this activity, you will compare productivity and respiration in three ecosystems.

✔**MATERIALS:**   calculator (optional)

☺**TIME:**   1/2 hour for activity, 1 hour for followup

✿**ACTIVITY:**

Calculate the missing values in the table, "Productivity and Respiration in Three Ecosystems" on the next page:

Net Primary Productivity (NPP) = Gross Primary Productivity (P) – Plant Respiration (RP)

Net Ecosystem Productivity (NEP) = NPP – Consumer Respiration (RC)

Net Ecosystem Respiration (R) = RP + RC

P/R = ?

Productivity and Respiration in Three Ecosystems
kilocalories/meter^2/year

Productivity and Respiration	Alfalfa Field	Short-Grass Prairie	Deciduous Forest
Gross Primary Productivity (P)	24,000	5,230	27,976
Plant Respiration (RP)	9,200	1,778	18,200
Net Primary Productivity (NPP)			
Consumer Respiration (RC)	800	2,379	9,172
Net Ecosystem Productivity (NEP)			
Net Ecosystem Respiration (R)			
P/R			

Source: Coleman, D. C., & P. F. Hendrix, Agroecosystems Processes. In Pomeroy, L. R., & J. J. Alberts (eds.), 1988, *Concepts of Ecosystem Ecology*. Amsterdam: Springer-Verlag. Includes data from Odum, E. P., 1971, *Fundamentals of Ecology*, Philadelphia: Saunders. Adapted by permission.

✎FOLLOWUP:

1.   Which ecosystem has the greatest Gross Primary Productivity? the greatest Net Primary Productivity? the greatest Net Ecosystem Productivity?

2.   Which ecosystem has the highest ratio of Productivity (P) to Respiration (R)? Which has the lowest? What does a ratio of 1.00 mean?

3.   The forest and prairie ecosystems are unmanaged, whereas the alfalfa field is managed for agricultural purposes. On the basis of the data in the completed table, what are two differences between agricultural and unmanaged ecosystems?

4.   What source(s) of respiration has been omitted from the table? If this (these) factor(s) were taken into account, how would the P/R ratio be affected?

5.   How would the P/R ratio be affected by deforestation? burning of fossil fuels?

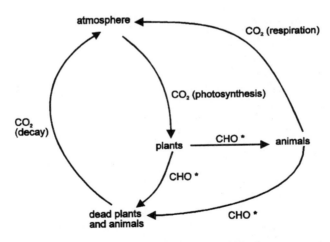

* CHO=Organic Carbon Compounds

Biological Carbon Cycle

Activity #18: Carbon Budget

The cycling of carbon through photosynthesis and respiration is only part of the global cycling of carbon. In addition to these biological processes, geochemical processes contribute to carbon cycling. The biological processes transfer carbon between living organisms and the environment; geochemical processes transfer carbon between sedimentary rocks and the atmosphere, oceans and living organisms. While biological processes can cause variations in atmospheric carbon dioxide over tens of thousands of years, geochemical ones work on a time scale of millions of years. Biological processes are, therefore, relatively short-term. Even more short-term, however, are human activities. Since the Industrial Revolution began 200 years ago, burning of fossil fuels has produced an amount of carbon dioxide equal to that which would be released if all life on Earth were immediately incinerated.

Carbon occurs primarily as carbon dioxide (CO_2) in air and water, organic carbon in living and dead organisms (proteins, carbohydrates, fats, and nucleic acids), and carbonate ions (CO_3^-) in water, rocks, shells, and bones. To understand how these are connected in a cycle, it is useful to think in terms of sources, sinks, and fluxes. Sources are carbon emitters; sinks are carbon absorbers; and, fluxes are flows of carbon between sources and sinks. A source may also be a sink. For example, the atmosphere is a source of carbon dioxide for photosynthesis, but it is a sink for carbon released during respiration, burning, and decay. Because carbon dioxide in the atmosphere is the biggest contributor to the greenhouse effect, scientists are concerned that continued increases in atmospheric carbon dioxide levels may lead to global warming.

☞In this activity, you will model the carbon reservoirs and fluxes, and consider what happens to the increasing carbon dioxide produced by human activities.

✔MATERIALS: calculator (optional)

☺TIME: 1 hour for activity, 1 hour for followup

❀ACTIVITY:

1.   Use the information in the table "Carbon Reservoirs" to complete the diagram, "Model of the Global Carbon Cycle" (see next page). Put the number of gigatonnes of carbon stored in each reservoir in the small boxes in each reservoir. You need to be aware that it is very difficult to make estimates of this kind. The values in the table are ones often cited and are useful for gaining understanding of the global carbon cycle. Research in this area is still going on, however, and there is much disagreement about the correct values among scientists. Each value has a degree of uncertainty, which has been omitted in order to simplify the activity. Gigatonnes (Gt)s represent the mass of the element carbon. One gigatonne equals 1000 million tonnes (10^9 tonnes) and 1 tonne equals 1000 kg.

Carbon Reservoirs

Reservoir	Carbon (Gts)
ocean surface	900
ocean life	3
organic material in ocean	1,000
deep ocean water	36,400
sediments	3,000
metamorphic/igneous rocks	20,000,000
sedimentary rocks	70,000,000
soil	1,600
fossil fuels	4,000
living land organisms	560
atmosphere	750

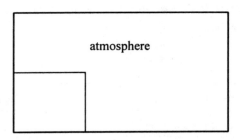

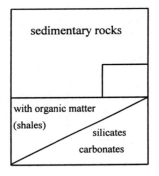

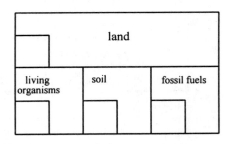

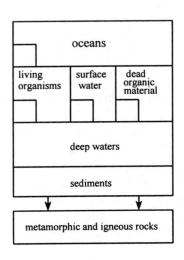

Model of the Global Carbon Cycle

2.   The fluxes of carbon between reservoirs is measured in gigatonnes of carbon per year (Gt/yr). Use the data below to calculate the net flux for the atmosphere, land, and oceans.

Carbon Fluxes

Direction of Movement	Flux (Gt/yr)
ocean to atmosphere	102
atmosphere to ocean	105
ocean surface to deep waters	39
deep waters to ocean surface	37
ocean surface to ocean life	28
ocean life to ocean surface	29
soil to atmosphere	60
life on land to soil	60
life on land to atmosphere	50
atmosphere to life on land	110
deforestation to atmosphere	1.6
fossil fuel combustion to atmosphere	5.4

3.   The average time that carbon atoms spend in a reservoir is called the **residence time**. You can calculate the residence time by dividing the number of gigatonnes of carbon in a reservoir by the total flux from that reservoir. For example, the total flux from the atmosphere is 102 (to the ocean) + 110 (to life on land) = 212 Gt/yr. The atmosphere contains 750 Gt of carbon. Dividing 750 by 212 gives a residence time of 3.5 years. Calculate the residence time for carbon in soil, land-based life, and ocean life.

✎FOLLOWUP:

1.   Which reservoir has the largest amount of carbon? What percentage of the total carbon in the land, ocean, and atmosphere (excluding the rocks of the Earth's crust) is in the atmosphere? Considering the answer, why is the level of carbon in the atmosphere considered so important?

2.   How much carbon is being added to the atmosphere each year? What are the sources of carbon dioxide responsible for the change? What process described in this activity is not cyclic? What process removes carbon dioxide from the atmosphere at the greatest rate?

3.   Scientists estimate that the carbon level of the atmosphere is increasing at a rate of about 3.2 Gt/yr and that the oceans absorb about 2.0 Gt/yr. How does this compare with the amount of carbon being added to the atmosphere each year? Develop several hypotheses to explain where the "missing" carbon could be going.

4.   Explain the differences in residence times between soil and life on land; life on land and life in the ocean.

5.   Discuss the significance of the relatively short residence time of carbon in the atmosphere.

Geochemical Processes in the Carbon Cycle

1. Carbon dioxide in air reacts with water to form carbonic acid.
2. Shales contain organic carbon, which reacts with oxygen in air to form carbon dioxide and enter the atmosphere.
3. Decaying organic matter in soil releases carbon dioxide, which reacts with water to form carbonic acid.
4. Rocks containing inorganic carbon (carbonates and silicates) react with carbonic acid, releasing bicarbonate and carbonate ions. These are carried by freshwater until they reach the ocean.
5. Marine organisms combine about half of the carbonate ions with calcium in the formation of shells and skeletons. When they die, these form sediments on the floor of the ocean.
6. The remainder of the carbonate ions form carbon dioxide, which enters the atmosphere.
7. Volcanoes and other types of eruptions release carbon dioxide into the atmosphere.

Activity #19: Litter Bugs

Traditionally, wildlife meant large animals, including birds and fish, hunted for sport. A more modern definition includes endangered species and animals important for tourism. As science has revealed the critical role of invertebrates and microorganisms in nature, people's concept of wildlife is coming to include these as well. Although inconspicuous, they are important in the natural recycling of matter, and affect an ecosystem's productivity.

Millipedes and other invertebrates break the leaves and other debris that fall to the forest floor into small bits and devour them. The fecal pellets they produce contain much undigested plant material, which becomes a food source for smaller animals, fungi, and bacteria. Meanwhile, carnivorous invertebrates are busy devouring others. The activities of this complex soil community convert insoluble nutrients to soluble ones that can be taken up by plants, to begin a new cycle.

☞In this activity, you will collect and identify small invertebrates from a sample of soil or leaf litter.

✔MATERIALS:

newspaper
small plastic bags
trowel or other digging implement
several small glass jars
bottle of rubbing alcohol
1 glass jar, at least 1 quart
funnel with wide stem (or cut off the top of a 2-
    liter plastic soda bottle and invert it)
piece of wire mesh to fit in funnel
paper towels
light source
masking tape
marking pen
magnifying lens
heavy rubber gloves (as needed)

⊘TIME: 3 hours for activity, 1 hour for followup

❀ACTIVITY:

1.   Collect samples of leaf litter and/or soil from a wooded area, compost heap, garden, lawn, or other area. If collecting in an area where there is trash or garbage, wear heavy protective gloves. Put the samples in plastic bags and use the masking tape and marking pen to label them. When you are finished collecting, wash your hands thoroughly with soap and warm water.

2.   Spread one sample at a time on newspaper and pick out the organisms you can see readily. Place them in the small jar with some of the alcohol.

3.   To collect organisms you may have missed, place the wire mesh in the funnel. Line the sides of the funnel with moist paper towls. Place the remaining soil and litter from the newspaper on top of the mesh. Pour a small amount of alcohol in the quart jar and put the funnel on top of the jar, with the stem in the jar. Arrange the light to shine on the sample. After a few days, small animals should fall into the alcohol.

4.   Sort, compare, and classify the organisms you have collected, using the magnifier if necessary. Consult field guides and other resources to identify the organisms. You may wish to make drawings or photos of them.

❧FOLLOWUP:

1.   Compare the organisms found in different samples you or your classmates have collected.

2.   Research the food habits of the organisms in one sample and make a food web for them.

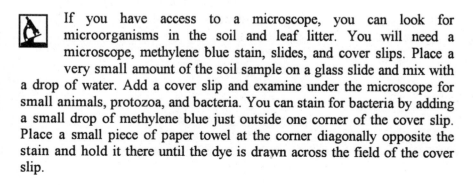

If you have access to a microscope, you can look for microorganisms in the soil and leaf litter. You will need a microscope, methylene blue stain, slides, and cover slips. Place a very small amount of the soil sample on a glass slide and mix with a drop of water. Add a cover slip and examine under the microscope for small animals, protozoa, and bacteria. You can stain for bacteria by adding a small drop of methylene blue just outside one corner of the cover slip. Place a small piece of paper towel at the corner diagonally opposite the stain and hold it there until the dye is drawn across the field of the cover slip.

Activity #20: A Lot of Rot

Composting and recycling are ways of reducing waste, but two-thirds of the garbage in the United States ends up in landfills. Dumping garbage is an ancient practice; archeologists often locate sites of ancient civilizations by middens—mounds of debris discarded many years ago. Out of a concern for public health, dumps have been replaced by sanitary landfills. Regulations require that they must have liners, to prevent toxic or unsanitary materials from reaching groundwater, and systems for controlling gaseous emissions.

Although modern methods help to contain materials in landfills and protect the environment, they slow the natural processes of decomposition. Material is compacted, preventing oxygen and water from reaching it. Recyclable organic materials can remain in good condition for 10 to 20 years. Newspapers have been found that are legible 35 years after being landfilled. The volume of material, therefore, decreases only slowly, a situation made even worse when nonrecylable materials are present.

As the population has grown, the amount of garbage produced has increased, but the land available for disposing of it has decreased. Cities, towns, and suburbs have to look elsewhere for sites for landfills, which often brings them into conflict with residents in other areas. Transporting solid wastes for great distances is expensive; it uses energy and contributes to air pollution as well. Incinerating materials before landfilling or after is controversial because of the possible release of toxic substances. For the present, the best solutions to the solid waste problem are to reduce waste, reuse materials, and recycle.

☞In this activity, you will explore how different conditions affect the decomposition of recyclable materials.

✔MATERIALS: banana peel
newspaper
5 self-closing small plastic bags
garden soil
water
masking tape or self-adhesive labels

⊘**TIME:**   2 hours for activity, 1 hour for followup

✿**ACTIVITY:**

1.   Cut the banana peel and newspaper into squares with 2-inch sides. Place one piece of each in each plastic bag.

2.   Label the bags, 1 through 5, and set them up as described below:

Bag 1: Seal the bag.

Bag 2: Add enough water to cover the peel and the newspaper. Squeeze out the air from the bag and seal it completely.

Bags 3, 4, 5: Add garden soil to cover the peel and the newspaper. Squeeze out the air and seal.

Bag 4: After sealing, poke small holes in the bag.

3.   Put bags 1, 2, 3, and 4 in a dark place. Put bag 5 in a sunny area or under a lamp. **Do not open the bags under any circumstances! Harmful bacteria or fungi may be present. When you have completed the activity, discard the bags unopened.**

4.   Check the bags each week until you observe signficant changes. Record your observations each time.

✎**FOLLOWUP:**

1.   What conclusions can you draw from your observations?

2.   Prepare a table comparing the potential uses and advantages and disadvantages of composting, recycling, and landfilling of solid wastes.

Activity #21: Succession

Changes occur in ecosystems over time. Stages of change can often be predicted on the basis of past observations and knowledge of the climate and geology of the region. A series of changes in the plants and animals living in an area is known as ecological succession. For example, ponds fill with sediment and become more shallow. As the area of open water decreases, new types of plants move in, gradually forming a bog. Eventually, the bog fills in completely and grasses and ferns appear. Trees that can tolerate wet soil begin to grow. As time goes on, the soil becomes drier and new species of trees move in and replace the earlier ones. If a fire destroys most of the trees and large shrubs, the area will revert to an earlier stage of succession until a new forest is established.

Succession may be observable over a few years, as after a forest fire, over many years, as after a farm is abandoned, or over many centuries, as after a glacier scrapes an area down to bare rock. One way to observe succession in a short time is to study examples of trees in various stages, from ones that have died but are still standing to those that have decayed completely.

☞In this activity, you will study succession by observing tree trunks in various stages of decay.

✔MATERIALS: magnifying lens
                  pocket knife

☉TIME: 2 hours for activity, 1 hour for followup

🍀ACTIVITY:

1.   Locate an area with trees in various stages, from standing dead trees to fallen and rotting logs. Select one species for which you can find at least one example of (1) a standing dead tree (2) a recently fallen tree (3) a fallen log with bark intact but rotting inside (4) a completely rotten log.

2.   Use the pocket knife, when necessary, to dig under the bark or into the wood. Use the magnifying lens to look for small organisms. For each of the four examples, make observations on the following:

> condition of the bark and the wood
> plant or fungal growths on the bark and/or in the wood
> moisture content of wood
> texture of wood (hard? soft? spongy?)
> animals under bark or in wood
> animals nesting in tree or log

❧FOLLOWUP:

1.   Describe the changes in the plant and animal life in the major stages of succession from a standing dead tree to a completely rotten log.

2.   What impact would the following have on succession in the area you studied? fire, logging, clearing of dead logs and brush, a large deer population (deer eat seedlings, but not ferns).

 If areas of woods with rotten logs are not accessible to you, you can study succession in a number of other habitats, two of which are described below.

Succession in an Abandoned Field

1.   Select three or more farm fields that have been abandoned at various times. If you cannot find out the years in which they were abandoned, you can at least estimate their relative age. The greater the size and number of trees, the greater the time that has elapsed since the field was last cultivated.

2.   In each field, choose a starting point at random (but not right at the edge) and place a stake at that point. Place a second stake at a distance of one meter. Tie a string between the two stakes and examine the plants along the string. Tabulate each new plant as you come to it, using the following categories: lichen, fungus, moss, fern, tree seedling, grass, small broad-leaved plant (stems not woody), shrub (woody plant with several to many stems starting at or near ground level), tree (woody plant with one main trunk). Extend your line for another meter and repeat the procedure. Continue until you have collected data for ten meters.

3.   Prepare a bar graph showing the frequency of each category of plant in each field. Place the names of the plant categories on the horizontal axis and the number of occurrences on the vertical axis. Use one bar for each field in each category; thus, if you sampled three fields, you will have three bars for each plant category. Use the graphs to compare the fields and to describe the stages in succession in an abandoned field. What factors account for the particular changes in plant life over time?

Succession in a Burn Area

1.   If you live near an area that has been recently damaged by fire, you can study succession as the area recovers from the fire. In this case, however, you will need to make observations on the burn area over a period of several months or a year. You can speculate on later stages of succession by observing the plants typical of nearby areas that escaped the fire.

2.   Select an area approximately one meter square in the middle of the burn area, using a square made of four meter sticks or a hula hoop. Observe and tabulate the plants found in the area using the categories described in step #2 for "Succession in an Abandoned Field." Repeat observations every month for as long as you can.

3.   Prepare a table in which you crosstabulate the types of plants with observation time. Make one column for each plant category and one row for each observation. Record the frequency of each plant category at each time. Use the frequency table to describe changes in succession in the burn area.

     Succession in a Micro-Ecosystem

You can use pond water to study succession. Follow the instructions for "Creating a Micro-Ecosystem" in Activity #13. When you have completed your observations, make a graph showing changes in the population over time.

ENERGY

Activity #22: Sunlight Becomes You

Energy flows through an ecosystem from an outside nonliving energy source through a food chain of living organisms. A food chain begins with producers, organisms that can capture the energy and convert it into living material. Organisms that cannot produce living material directly from a nonliving energy source, the consumers, depend upon the producers for food. At each link in the food chain, some usable energy is converted into nonusable energy, usually in the form of heat. The more links there are in the food chain, the less usable energy remains at the top of the chain. Because the useful energy in the system is constantly being degraded into nonusable energy, an ecosystem requires a continual input of energy.

On Earth, there are two main types of ecosystems. One depends ultimately on energy from the sun, which is captured by green plants and algae and converted into living material. The other depends upon energy released by chemical reactions and producers that can convert that chemical energy into living matter. The first type of ecosystem is the one with which we are most familiar and the one that supplies the food we eat. The second type of ecosystem is found among certain bacteria near the surface of the Earth and in communities of organisms found around the deep ocean vents. Recently, some scientists have found evidence that this second ecosystem may be much more extensive than previously thought. They suggest it may extend several miles under the Earth's surface and contain as much or more living material than the one with which we are most familiar.

Each level in a food chain is called a trophic level. The rate at which organic matter (biomass) accumulates at each trophic level is a measure of the productivity of the ecosystem at that level. The productivity of the producers in an ecosystem is known as primary productivity, whereas that of the consumers is known as secondary productivity. Productivity can also be classified as gross productivity (before the organisms use the biomass to release energy through respiration) and net productivity (what remains after the organisms have used the biomass to produce energy for their life activities).

☞In this activity, you will calculate the amount of light energy from the sun required to produce the food you eat in one day.

You may wish to do this activity in conjunction with Activity #31.

✔**MATERIALS:**   calorie chart

☾**TIME:**   1 hour for activity, 1 hour for followup

❀**ACTIVITY**

1.   Keep track of the food that you eat during one day. Use two columns, one for animal products (meats and dairy foods) and one for plant products (vegetables, fruits, and foods derived from them, such as vegetable oils and margerine).

2.   From the calorie chart and/or information on the food packages, estimate the number of calories, actually kilocalories (kcal), of animal food (C_A) and plant food (C_P) you have eaten.

3.   Estimate the number of calories of plant food eaten by the animals you ate (C_{AP}). (Assume that animals, such as cows and chickens, are 10% efficient in processing food, i.e., only 90% of the plant material they eat can be passed on to you as food. The rest is used up in the daily activities of the animal or dissipated as heat.)

$$C_{AP} \text{ (kcal)} = C_A \text{ (kcal)} \div 0.10.$$

4.   Find the total number of calories of plant material (C_T) that you ate directly or indirectly.

$$C_T \text{ (kcal)} = C_{AP} \text{ (kcal)} + C_P\text{(kcal)}.$$

5.   Calculate the amount of light energy, in kilocalories, (C_L) required to produce the plant material you ate directly or indirectly. (Assume that plants are only 1% efficient in converting sunlight to stored organic matter.)

$$C_L \text{ (kcal)} = C_T \text{ (kcal)} \div 0.01.$$

✎**FOLLOWUP:**

1.   Make a food chain linking you to the foods you ate directly and indirectly for the day.

2.   Draw an energy pyramid showing, from top to bottom, the calories of animal food you ate (C_A), the calories of plant food the animals ate (C_{AP}), and the calories of light energy needed to produce the plants eaten by the animals ($C_{AP} \div 0.01$).

3.   Identify the quantities that represent each of the following: primary productivity, secondary productivity, gross productivity, net productivity.

4.   Which is a more efficient use of the sun's energy, eating plant food or eating animal food? Explain your answer.

5.   Explain what the phrase, "eating lower on the food chain" means.

6.   How could eating lower on the food chain contribute to solving the problem of food shortages in developing areas of the world?

Activity # 23: Fossil Fuels

The economies of the industrialized world run on fossil fuel. Coal, gas, and petroleum, formed hundreds of millions of years ago by decaying plants and animals, have provided modern people with a supply of stored energy from the sun. Fossil fuels have allowed us to move from a society based primarily on energy from people and living plants and animals to one based on fossil fuels. Special conditions that existed when coal, gas, and petroleum formed are not present now, so they can no longer form in significant amounts, if at all. Furthermore, formation of fossil fuels is a very slow process, too slow for any replacement to keep step with current use. Because fossil fuel reserves are limited and cannot be replaced, they are said to be **nonrenewable**.

Limited supplies are, however, not the only concerns. When fossil fuels are burned, they produce carbon dioxide, the principal contributor to the greenhouse effect. Increases in atmospheric levels of carbon dioxide have been observed during this century and threaten to cause global warming. Other gases emitted by fossil fuels contribute to air pollution and acid rain. Coal mining, particularly of the above-ground type called strip mining, damages the landscape, while dust and noxious gases in underground mines are a health hazard for miners. Finally, many people are concerned that our dependence on oil, mostly from the Middle East, makes us vulnerable to the politics of that area.

☞In this activity you will estimate the amount of fossil fuel you consume directly, through transportation, electrical appliances, and home heating, and indirectly.

✔**MATERIALS:** automobile (If you do not own a car, obtain information from someone who does. If you use public transportation, you may be able to obtain information from the transport company. If you walk or ride a bicycle, transportation will not contribute to your consumption of fossil fuel.) utility bills

☺**TIME:** 2 hours for activity, 1 hour for followup

✿ACTIVITY:

1.   To estimate the amount of fossil fuel you consume in transportation:

    a.   divide the number of miles you drive in one year by the number of miles per gallon your car gets to find the number of gallons of fuel you use in a year. If you do not know this, you can keep a record of fuel consumption for one week while completing step #5.

    b.   multiply the number of gallons of fuel you use per year by 125,000 BTUs (the amount of energy in one gallon of gasoline) to find the number of BTUs you use in one year.

2.   To estimate the amount of fossil fuel equivalent to the energy you consume in electrical appliances:

    a.   use the table on "Energy Requirements of Household Electric Appliances" on page 96 to find the appliances you use.

    ⬌   You may want to save this information to use in conjunction with Activity #25.

    b.   add the values for each appliance to find your annual energy consumption for electrical appliances in kilowatt-hours per year (kWh/yr).

    c.   multiply the number of kilowatt-hours per year by 3411 (the number of BTUs equivalent to 1 kilowatt-hour) to find the number of BTUs of electrical energy you use annually.

    Note: The electricity you receive in your home may not be generated by burning fossil fuel, but by some alternative, such as hydroelectric or wind power. If that is the case, you can omit this value from the estimate of your fossil fuel use. However, it may be of interest to you to calculate the equivalent amount of fossil fuel that would be required if your utility company did use fossil fuel in producing electricity.

3.   To estimate the amount of fossil fuel you consume in home heating:

a.   If your home is heated by electricity, you can obtain a monthly heating bill and convert the kilowatt-hours to BTUs per year, as in step #2.

b.   If your home is heated by natural gas, oil, or wood, you can use your monthly heating bill and the following information to convert your fuel use into BTUs per year:

1 cubic foot of natural gas	=	1,031 BTUs
100 pounds of coal	=	1,111,100 BTUs
1 gallon of fuel oil	=	140,000 BTUs
120 pounds dry wood	=	948,000 BTUs.

4.   Add the number of BTUs from steps #1, #2, and #3 to find your total fossil fuel consumption per year.

5.   Keep track of the purpose, time spent, and length of your car trips for one week. Use the following categories to record the purpose: home to work, work related, family business, visiting friends, shopping, civic or religious, vacation, doctor or dentist, and pleasure. If you record length in miles, multiply by 1.6 to convert to kilometers.

✎FOLLOWUP:

1.   The figure you calculated is the energy you use **directly**. It does not include the energy you use **indirectly**. Indirect energy use includes energy used in manufacturing products you buy, growing and processing food, and transporting food and products to you. Approximately 75% of the energy we use is used indirectly. Therefore, you need to multiply the total number of BTUs you calculated by 4 to obtain your total energy consumption. Compare this to the average total energy consumption per person in the United States of 300 million BTUs.

2.   Prepare a bar graph of your car trips to show the percentage of trips in each of the categories in step #5 above. (These are listed in order from most to least frequent purposes of motor vehicle trips in the United States). How do your results compare with the national results? How could you conserve gasoline?

3.   Each gallon of gasoline burned produces 24 pounds of carbon dioxide. Carbon dioxide is the prinicpal contributor to the greenhouse effect. Calculate the number of pounds of carbon dioxide your car produces each year. This is not your total, as all other energy you use (except for

solar, wind, and nuclear) also produces carbon dioxide. The production of carbon dioxide per person in the United States is approximately 5 tons per year.

Energy Requirements of Household Electric Appliances
(kilowatt-hours consumed annually)

Air cleaner-216
Air conditioner-860 (based on 1000
    hours of operation per year)
Blanket-147
Blender-1
Broiler-85
Clock-17
Clothes dryer-993
Clothes washer-103
Coffee maker-140
Dehumidifier-377
Dishwasher-165
Fan (circulating) -43
Fan (furnace)-650
Fan (window)-200
Freezer,16.5 cu.ft , automatic
    defrost-1,820
Freezer, 16 cu. ft., manual defrost-
    1,050
Frying pan-100
Hair dryer-25
Heat lamp-13
Heating pad-10
Hot plate-90

Humidifier-163
Iron-60
Microwave-100
Mixer-2
Radio-86
Radio + Record player-109
Range + oven-596
Refrigerator, automatic defrost,
    17.5 cu. ft.-1,591
Refrigerator, manual defrost, 12.5
    cu. ft.-1,500
Sewing machine-11
Shaver-.5
Sun lamp-16
Television, black/white-100
Television, color-320
Toaster-39
Toothbrush-1
Vacuum cleaner-46
Waffle iron-20
Waste disposer-7
Water heater-4,219
Water heater, quick recovery-4,811

Source: Edison Electric Institute, 701 Pennsylvania Avenue, NW, Washington, DC 20004-2696.

Ways to Reduce Carbon Emissions*

What You Can Do	Reduction
Tune up your car	265 lbs
Drive a car with 30 mpg instead of 20 mpg	880 lbs
Drive a car with 40 mpg	1,320 lbs
Drive a car with 50 mpg	1,580 lbs
Take a train rather than fly	10 lbs/100 miles
Carpool with five others	5,000 lbs

*Based on driving 10,000 miles per year.

Source: From *The Greenhouse Trap* by World Resources Institute. Copyright © 1990 by World Resources Instititute. Reprinted by permission of Beacon Press.

Ways to Save Energy*

What You Can Do	Energy Savings
Improve insulation in your hot water heater	300 kWh/yr
Switch from resistance heater to heat pump	2,000 kWh/yr
Switch from typical refrigerator/freezer to more efficient model	1,250 kWh/yr
Update central air conditioning	1,000 kWh/yr
Substitute an 18-watt compact fluorescent bulb for a regular 75-watt bulb for 8 hours	170 kWh/yr

*Every kilowatt-hour saved reduces carbon emissions by 0.4 lbs.

Source: From *The Greenhouse Trap* by World Resources Institute. Copyright © 1990 by World Resources Institute. Reprinted by permission of Beacon Press.

Activity #24: Energy and Agriculture

Traditional agriculture uses sunlight to store energy in plants through the process of photosynthesis. Animals, including humans, obtain their energy from that stored in the plants. Traditional agriculture is said to be "labor intensive" because, except for the energy of the sun used directly in photosynthesis, it depends solely on the labor of people and animals. Modern, mechanized agriculture, on the other hand, uses stored energy in the form of fossil fuels to run machines for plowing, cultivating, irrigating, harvesting, and transporting of crops, as well as for fertilizers and pesticides, which require the use of fossil fuels in their manufacture. Modern agriculture is thus said to be "energy intensive." The drawing at the top of the next page illustrates the evolution of agriculture.

Farms can be thought of as special types of ecosystems, or agroecosystems, which differ in a number of significant ways from natural ecosystems. For example, people try to prevent succession, or change over time, in agroecosystems so that they will continue to produce the same crops year after year. Farmers control competition by ridding farm areas of weeds; they control predation by spraying insects and other pests. Farming also reduces diversity, by planting the same crop over wide areas of the land, a practice known as monoculture. Plowing is an unnatural disturbance of soil that can lead to erosion, and chemical fertilizers can destroy the physical and chemical makeup of soil. In recent years, scientists and farmers have begun to work together to develop a modern form of agriculture that is more in tune with the principles of ecology and will, in the long run, provide a more sustainable yield.

☞In this activity, you will compare the flow of energy on a traditional farm with that on a modern mechanized one.

⊘TIME: 1 1/2 hours for activity, 1/2 hour for followup

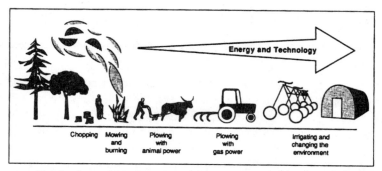

Evolution of Human Activities and the Agroecosystem

Source: C. M. Ghersa, M L. Roush, S. R. Radosevich, & S. M. Cordray, 1994 (Feb.), Coevolution of Agroecosystems and Weed Management, *Bioscience*, *44*(2): 85-94. Copyright ©1994 by the American Institute of Biolgical Sciences. Reprinted by permission.

❀ACTIVITY:

Study the list of symbols below and the way they are used to represent energy flow in a home garden plot in the diagram "Garden Energy Flow Chart" (see page 102). When you feel that you understand the diagram, prepare a similar one for a traditional farm and for a modern, mechanized farm. In the case of the traditional farm, assume that the farmer and his family eat most of the crop and carry the rest to a nearby village to sell. The energy expended by the farmer (and his family) is roughly 6×10^4 kcal/acre/year, and the energy yield of the crop is 1.8×10^6 kcal/acre/year. In the case of the modern farm, assume that the farmer raises corn as a cash crop and ships almost all of it to the city for sale. The energy expended by the farmer (and his family and hired help) is 1.2×10^4 kcal/acre/year, the energy of the yield is 20×10^6 kcal/acre/year. The modern farmer uses 7.1×10^6 kcal/acre/year of fossil fuel.

outside energy source

stored energy

heat loss

work performed

photosynthesis by green plants

consumer

✎FOLLOWUP:

1.   In addition to human energy and energy from fossil fuels, both types of agriculture have an unlimited energy source. What is it?

2.   What are the sources of stored energy for each type of agriculture?

3.   Which type of farm produces the higher yield? How many times greater is it than the smaller yield?

4.   Which type of farm requires a greater amount of human energy? How many times greater is it than for the other type?

5.   Calculate the energy output for each type of farm for each kilocalorie of energy input. Which type of farm is more efficient (i.e., has the higher ratio)?

6.   What reasons can you think of for the continued growth of mechanized farming?

Garden Energy Flow Chart

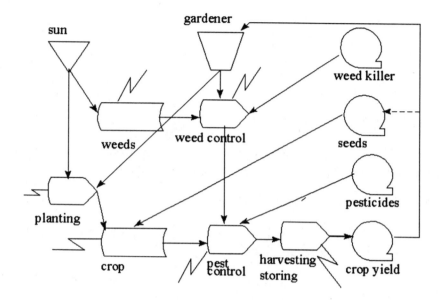

Activity #25 : Wind Power

People have been using wind power since ancient times. Wind patterns are so predictable that sailors used them for telling direction, before there were magnetic compasses, as well as for locomotion. Windmills were commonly used to pump water and thresh grain until hydroelectric and fossil-fuel based electric generating plants replaced them. Conventional electric energy generation consumes over 40% of the fossil fuels used in the United States. Concerns about limited resources and environmental effects of fossil fuel combustion have stimulated renewed interest in clean, renewable energy sources, such as wind, solar, geothermal, and tidal power.

World wind energy generating capacity increased from 15 megawatts (mW) in 1981 to over 2,500 mW in 1992. Modern wind turbines could provide at least 20% of the electricity used in the United States, but growth in wind generating capacity has slowed in the 1990s, due to lack of economic incentives. To generate this much electricity would require 47 million hectares (18,000 square miles), an area 1/7 that of the state of Texas. California, which generates 1% of its electrical energy needs by wind power, leads the nation; but other areas of the country, such as New York state, have even greater potential wind power (see map on next page).

Winds of 3.6 meters per second (m/s; 8 mph) can be used to generate electricity, although this is most economical where wind velocities are over 6.7 m/s (15 mph). Flat terrain, with less than a 3% grade, and no large features to break the wind for a distance of 1 mile, is ideal for a wind generating site. Because of friction between the moving air and land, wind velocities increase with altitude. A wind of 4.4 m/s (10 mph) at 9.1 m (30 ft.) will move at 5.2 m/s (11.7 mph) at 24.4 m (80 ft). Wind power increases with the cube of wind velocity; thus, a small increase in velocity produces a relatively significant increase in power. Installation costs for small wind turbines are approximately $1,000 per kilowatt (kW); operating and maintenance costs are between 1¢ and 2¢ per kilowatt-hour (kWh).

☞In this activity, you will evaluate sites in your area to determine whether it would be feasible and economical to use wind power to generate electricity for your home.

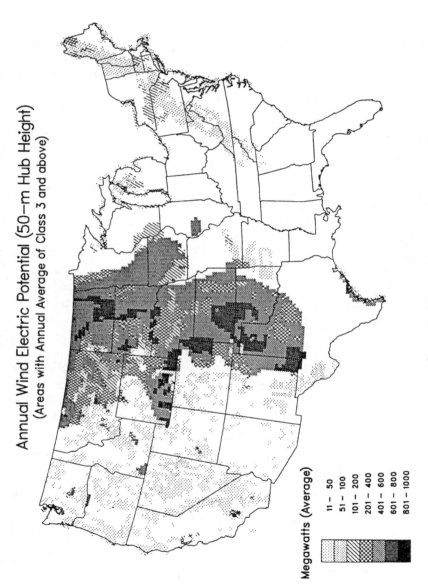

Annual Wind Electric Potential (50–m Hub Height)
(Areas with Annual Average of Class 3 and above)

Megawatts (Average)

11 – 50
51 – 100
101 – 200
201 – 400
401 – 600
601 – 800
801 – 1000

Source: Battelle Pacific Northwest Laboratories. Reprinted by permission.

✔**MATERIALS:**   ping pong ball
30 cm monofilament fishing line
protractor
small bubble level
large needle (length: longer than diameter of ping
 pong ball; diameter of eye: large enough
 so fishing line can be threaded)
red marking pen
glue

☾**TIME:**   4 hours for activity; 1 hour for followup

✿**ACTIVITY:**

1.   **Making a wind indicator:** Color the fishing line with the marker. Thread the fishing line through the eye of the needle and push the needle through the ping pong ball. Remove the needle and knot the end on the far side of the ping pong ball. Glue the knot to the ping pong ball and the other end of the line to the center of the straight side of the protractor, as shown in the drawing. Glue the level to the protractor.

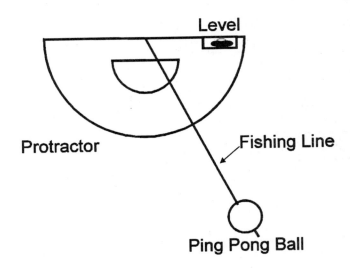

2.   To use the wind indicator: Hold the indicator so that the bubble is level and faces into the wind. Record the angle at which the fishing line crosses the protractor. Use the table "Converting Angles to Wind Velocity" to find the velocity of the wind.

3.   Select three level areas that you think would make good sites for a wind generator. Measure the angle at each site at the following times: 8 a.m., 12 noon, 4 p.m., and 8 p.m. Average the four readings for each site. For each site, select the angle from the table closest to the average angle, and convert angle to wind velocity.

Converting Angles to Wind Velocity

Angle	Wind Velocity (m/s)
90.0°	0.0
87.5°	2.0
82.5°	3.0
77.5°	4.0
75.0°	4.5
70.0°	5.0
67.5°	5.5
65.0°	6.0
62.5°	6.5
60.0°	7.0
57.5°	7.5
53.5°	8.0
50.0°	8.5
47.5°	9.0
45.0°	9.5
42.5°	You probably should not be outdoors!

Source: Adapted from New York Energy Education Project, 1988, *Renewable Energy: Student Activities* (revised edition), Albany, NY: The Energy Project.

Estimated Annual Energy Output at Hub Height
(thousand kilowatt-hours per year)

Average Speed (m)	Rotor Diameter (m)							
	1	1.5	2	3	4	5	6	7
4.0	0.1	0.3	0.6	1.3	2.3	3.9	5.2	7.1
4.5	0.2	0.5	0.8	1.9	3.4	5.3	7.6	10
5.0	0.3	0.6	1.0	2.3	4.1	6.5	9.3	13
5.5	0.3	0.7	1.3	2.9	5.2	8.2	12	16
6.0	0.4	0.8	1.4	3.3	5.8	9.0	13	18
6.5	0.4	0.9	1.7	3.8	6.7	10	15	20
7.0	0.4	1.0	1.8	4.0	7.0	11	16	22
7.5	0.5	1.1	2.0	4.6	8.1	13	18	25
8.0	0.5	1.1	2.0	4.5	7.9	12	18	24
8.5	0.6	1.3	2.4	5.3	9.5	15	21	29
9.0	0.7	1.6	2.8	6.3	11	18	25	34

Note: Efficiency decreases from about 30% to about 10% as rotor diameter increases, but decrease is not regular. Efficiencies used to calculate table were based on published data.

Source: Copyright © 1993 by Paul Gipe. Reprinted from *Wind Power for Home & Business: Renewable Energies for the 1990s and Beyond,* with permission from Chelsea Green Publishing Company.

4.   You can check your wind velocity readings in two ways: the Beaufort Wind Scale (see page 108), based on observing how smoke, leaves, and flags move, and the Griggs-Putnam Index (see page 109), based on the long-term effects of prevailing winds on the shapes of trees.

✎FOLLOWUP:

1.   How well suited are the sites you studied for using wind power to generate electricity? Explain the basis for your conclusion.

2.   Add the kilowatt-hours for each appliance in your household (see table of "Energy Requirements of Household Electric Appliances" on page 96) to find the total anuual electrical energy use for your household. Or, you can obtain the actual number of kilowatt-hours used from your monthly bills and multiply by 12.

3.   Use the table, "Estimated Annual Energy Output" (see page 107) to estimate the number and size of wind generators you would need to provide electricity to your household.

4.   Use the energy from step #2 and information in this activity to estimate the cost of using wind power.Would wind power be more or less economical as a source of electricity than your present source? (If you don't know the actual cost, ask you electric company for information on cost per kilowatt-hour.)

Beaufort Wind Scale

Beaufort #	Observation	Wind Velocity km/h (mph)
0	smokes rises straight up	0-1.6 (0-1)
1	smoke drifts slowly	3.2-4.8 (2-3)
2	leaves rustle	6.4-11.3 (4-7)
3	twigs move, flags extended	12.9-19.3 (8-12)
4	branches move, dust and paper rise	21.0-29.0 (13-18)
5	small trees sway	30.6-38.6 (19-24)
6	large branches sway, wires whistle	40.3-49.9 (25-31)
7	trees move, walking is difficult	51.5-61.2 (32-38)
8	twigs break off trees	62.8-74.1 (39-46)
9	branches break, roofs damaged	75.7-87.0 (47-54)

Mean Annual Wind Speed by the Griggs-Putnam Index

Observations	Mean Annual Wind Speed km/h (mph)
trunks vertical; branches and twigs not bent downwind	0-9.7 (0-6)
branches and twigs slighlty bent downwind; upwind branches short or have been stripped away	11.3-14.5 (7-9)
trees slightly deformed; branches and twigs as above; branches stream downwind slightly	14.5-17.7 (9-11)
effects on upwind branches more pronounced; branches stream downwind more definitely	17.7-21.0 (11-13)
branches stream downwind to the extent that they appear on one side of the trunk only	21.0-25.8 (13-16)
main trunk is predominantly vertical, but trunk and branches bend away from the prevailing wind	24.2-29.0 (15-18)
trunk is bent almost horizontally away from the prevailing wind	25.8-33.8 (16-21)

Activity #26: Solar Still

The sun is the ultimate source of almost all of the Earth's energy. People obtain energy from the sun when they eat plants, which have captured the sun's energy in the process of photosynthesis. The animals they eat or use to do work also depend on plants for their energy. Living plants, as well as fossil fuels, which are the remains of organisms that lived hundreds of millions of years ago, are used for fuel. Water used in generating electricity depends ultimately on the water cycle, which is powered by the sun. Uneven heating of air and water by the sun creates winds and ocean currents. Wind power, as you learned in the last activity, is a growing alternative energy source, and temperature differences in the oceans are used as an energy source in some pilot projects. All of these, however, capture only a small part of the energy that reaches Earth from the sun. Over 40,000 quadrillion BTUs (quads) of sunlight reach the continental United States each year, compared to the current total energy usage of 80 quads.

Solar energy is used for generating electricity and for heating buildings and water. Photovoltaic cells and films, which are made of materials that generate electrons when struck by light, increased by 15% a year, on average, in the 1980s. Although prices for solar electric power have fallen to about 10% of what they were when first introduced in the 1970s, the cost per kilowatt-hour is still about 6 times that of electricity generated by fossil fuels. In addition to its availability, solar energy has the advantage of producing little or no pollution. But because it is a diffuse energy source, it requires large collecting areas; to supply the energy needs of one person per day requires 40 square meters (430 square feet).

Heating of residential and commercial buildings by solar energy can be active, using devices that collect and store heat from sunlight, or passive, using the orientation and structure of buildings to take maximum advantage of solar energy. Passive methods have been used since ancient times. The Anasazi of the American southwest, for example, built their cliff dwellings so they received maximum warmth from the sun in winter and cooling in the summer. Solar energy can be used also for desalinization (producing freshwater from ocean water), although most current systems use electrical energy.

☞In this activity, you will construct a solar still to provide enough water for one person for one day.

✔**MATERIALS:**   according to the design of your solar still
salt
water
measuring container marked in milliliters
scale measuring in grams, or:
> 1 12-inch ruler
> 1 pencil
> 2 paper clips
> aluminum foil

☻**TIME:**   will vary

✿**ACTIVITY:**

On February 4, 1982, six days into a solo sailing trip from the Canary Islands to the Caribbean, Steven Callahan's boat sank. For the next 76 days and 1800 miles, he drifted in a 5 1/2-foot-long inflatable raft, until he reached the islands of Guadeloupe. At the time of the shipwreck, Steven had two army surplus solar stills on board, one of which worked a little, the other not at all. Using some of the few items he had been able to salvage from the boat, he improvised a solar still from a plastic box with lid, three cans, black cloth, plastic bags, and tape. He put the cans in the box, and put pieces of the cloth, which he had soaked in sea water, in the bottom of each can. Then, he propped up the lid of the box and made a plastic tent over the opening, which he secured with tape. Unfortunately, the still did not work because he could not seal the plastic to the box tightly enough. Luckily, he was then able to repair the army surplus solar still; between water from that still, rainwater, and the fish he caught and ate, he was able to avoid dying of thirst.

Can you improve on Steven's design and build a solar still that produces enough freshwater from salt water to sustain a person (473 milliliters or 1 pint a day)?

1.   Design and build a still, using materials that are readily available, that will produce 473 ml or 1 pint of water a day.

2.   Prepare a saltwater solution by dissolving 1.5 grams of salt in 45 ml of water. If you do not have a scale or balance calibrated in grams, you can make a crude balance using a 12-inch ruler, a pencil, aluminum foil, and

some paper clips. Balance the ruler on the pencil so that the ruler is parallel to the top of a table. Cut out two squares of aluminum foil of exactly the same size. Form a cup out of each of them. Place one on each end of the ruler, and again balance the ruler. Put 1 1/2 paper clips in one cup. Add salt to the other cup until the ruler is balanced again. (How can you cut a paper clip exactly in half?)

3.   When you have a working still, measure the number of milliliters of salt water you need to add to produce 473 ml of freshwater.

✎FOLLOWUP:

1.   What was the purpose of the black cloth? (You may want to review Activity #2.)

2.   It takes 540 calories of heat energy to convert 1 gram of water to water vapor. A milliliter of pure water has a mass of 1 gram. Although the mass of a milliliter of salt water would be somewhat larger, use this conversion to obtain an approximate value for the number of calories of sunlight needed to produce 473 ml of freshwater.

3.   What natural cycle on Earth is similar to the processes going on in the solar still? Explain.

4.   It requires 2,000 square centimeters (2 square feet) of water surface to produce 1 liter (0.26 gallons) of freshwater. Assuming that the average person in the United States uses 340 liters of water, how much area would be required to produce that amount of freshwater from salt water?

5.   Currently, it is more economical to use reverse osmosis and electrical energy for desalinization, rather than distillation and solar energy. It requires between 3.5 kilowatt-hours and 9.0 kilowatt-hours of electrical energy to produce 1000 liters of freshwater. Using the rates you are charged for electrical energy and for water, calculate how the cost of desalinized water obtained by the reverse osmosis process would compare with your current water cost.

Activity #27: Radioactive Decay[1]

Radioactive materials are forms of chemical elements that are unstable and change spontaneously until they form a stable isotope. The changes that a radioactive material undergoes is called a radioactive series, which may consist of one to many steps, and the process is known as radioactive decay. During many of the changes from one isotope to another, radiation is emitted in the form of nuclei of helium atoms (alpha radiation), electrons (beta radiation), or X rays (gamma radiation). Radiation of any type can be harmful, but the degree of danger depends upon the type of radioactive isotope, the radiation it emits, how long it remains radioactive, and how it is metabolized by living organisms.

The rate at which radioactive isotopes decay to stable ones depends upon the isotope and may vary from fractions of a second to millions of years. The time it takes for half of the initial material to form a stable isotope is called the half-life of the isotope. Although it is possible to predict the path of radioactive decay and the half-life, it is impossible to predict that any one atom of a radioactive isotope will decay at a particular time. The fate of any one atom in a mass of radioactive material is thus a matter of chance, that is, scientists can give only the probability that the atom will decay. A graph of the amount of radioactive material remaining over time follows a curve known as an exponential decrease; it can be described by a mathematical formula involving an exponent. An exponential decrease is characteristic not only of radioactive decay, but also of many other processes involving chance.

The curve of radioactive decay can be thought of as the inverse of the exponential increase of populations, which can be characterized by the doubling time. The doubling time of populations is determined by the birth and death rates; because these can vary over time, the doubling time need not be constant. In radioactive decay, however, the birth rate of new isotopes in the series and the death rate of the preceding ones is constant; thus, the half-life for an isotope is constant.

[1] Adapted from *Energy from Nuclear Reactions. Teaching About Energy, Part Four* by John Lord. Copyright © 1987 by Enterprise for Education, Inc. Reprinted by permission.

☞In this activity, you will simulate the decay of a radioactive substance, plot the decay curve, and determine the half-life.

✔**MATERIALS:**   popcorn kernels
a square box with lid (A flat box is best for this activity. Although a small box is most convenient, it may be difficult to find one. If you have trouble finding a square box, use a clean pizza box.)
graph paper

☉**TIME:**   1/2 hour for activity, 1/2 hour for followup

❀**ACTIVITY:**

1.   Remove the lid from the box and make an "X" inside the box on one side only.

2.   Count 100 popcorn kernels and place them in the box.

3.   Put the lid on the box and shake it. Take the lid off and remove all of the kernels that have their pointed ends pointing toward the side you marked with an "X." If any are pointing directly to one of the corners of that side, remove half of them. Do not put the removed kernels back in the box. Record the number you removed and the number left in the box in the data table, "Half-Life of Popcorn Kernels," on page 117.

4.   Repeat step #3 until you have completed the data table. Make a graph of the data, with the trial numbers on the horizontal axis and the number of kernels remaining on the vertical axis.

✎**FOLLOWUP:**

1.   What is the shape of the graph? What is this type of curve called?

2.   Is it important that the box be exactly square? What would your graph look like if the box were rectangular, but not square?

3.   Why does this produce the same type of curve as radioactive decay?

4.   What is the half-life of the popcorn kernels? (In the case of popcorn kernels, the half-life is the number of trials it takes to reduce original number of 100 to 50.)

Half-Life of Popcorn Kernels

Trial #	# Kernels Started With	# Removed	# Remaining
1			
2			
3			
4			
5			
6			
7			
8			
9			
10			

Half-Life of Radioactive Isotopes

Isotope	Half-Life
lawrencium-257	8 seconds
francium-223	21 minutes
iodine-131	8.1 days
strontium-90	28 years
carbon-14	5,568 years
uranium-238	4.5 billion years

Activity #28: Radiation Exposure[1]

The amount of radiation that is absorbed by an exposed organism is measured in units called the radiation adsorbed dose, or rad. The damage to the living organism depends upon the type of radiation (see box on page 120) received and the extent to which it can penetrate and injure the organism. The unit which measures the biological effects due to a radiation adsorbed dose is the rem. One rem equals 1000 millirem. A dose of about 500,000 mrem is lethal to 50% of the people (the lethal dose 50, or LD_{50}). Much lower doses of radiation, around 50,000 mrem, can cause perceptible physiological harm. The exposure limit established for workers in the United States is 5,000 mrem a year, while that for the general public is 200 mrem. Exposure to radiation is associated with certain types of cancer, as well as genetic and birth defects. Radiation comes from natural and human-made sources. The amount of radiation a person receives depends upon many factors, some of which you will explore in this activity.

☞In this activity, you will estimate your exposure to radiation from natural and human-made sources.

☺TIME: 1 hour for activity, 1/2 hour for followup

❀ACTIVITY:

Use the table, "Potential Sources of Radiation Exposure" (see pages 122 and 123) to estimate your exposure to radiation in millirem per year for each source. The table is divided into two parts: natural sources and human sources. When you have completed the table, add up all the numbers to find the total number of millirems.

---

[1] Adapted from *Energy from Nuclear Reactions. Teaching about Energy, Part Four* by John Lord. Copyright © 1987 by Enterprise for Education, Inc. Reprinted by permission. Data from National Council on Radiation Protection and Measurements, 1987, *Ionizing Radiation Exposure of the Population of the United States* (Report # 93), Bethesda, MD: National Council on Radiation Protection and Measurements.

Types of Radiation

Type	Components	Velocity	Penetration	Damage
alpha	2 neutrons + 2 protons	slowest	cannot penetrate skin	harmful if ingested or inhaled
beta	electrons	faster	can penetrate a few millimeters into skin	most serious if ingested or inhaled
gamma rays	waves	fastest	can penetrate deep into body	can damage internal organs even when not ingested or inhaled

Radiation Sources

United States Population

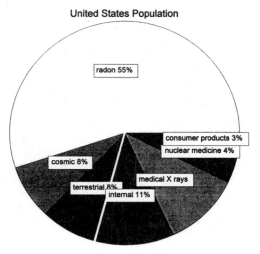

Source: National Council on Radiation Protection and Measurements, 1987.

✎FOLLOWUP:

1.   Until a few years ago, the annual maximum permissible exposure for the general public was 500 mrem. It is now set at 200 mrem. How does your radiation exposure compare with these values?

2.   Which contributes more to radiation exposure for the average American, natural or human-made sources? Which is the largest contributor to natural sources? to human-made sources?

3.   Which of the sources of radiation could be reduced or eliminated? Which could not? Explain.

Sources of Radiation

Natural Sources

   Cosmic rays from outer space,or from the radiation belts surrounding the Earth, bombard the atmosphere, bringing potentially dangerous radiation. Many of the rays are stopped by the atmosphere before they reach the Earth's surface, thereby reducing the potential harm. But people living at higher altitudes receive more radiation from cosmic rays than do people living at sea level. Traveling in airplanes at high altitudes has a similar effect, although the exposure time is short for all except airline employees.

   Radioactive isotopes occur naturally in the rocks and soils of the Earth's surface and these emit radioactivity. The walls of buildings provide some protection, but some radiation leaks through. Radon and its decay products are the most significant of these isotopes.

   Foods contain some radioactive materials because plants absorb them from the soil in which they grow or from the interaction of cosmic rays with molecules in the atmosphere. Potassium-40 is absorbed from the soil by plants when they take in the more common and stable isotope, potassium-39. Cosmic rays convert a fraction of stable nitrogen-14, which makes up most of our atmosphere, to carbon-14. When plants take in carbon dioxide, some of it is made with the radioactive carbon-14 instead of the stable isotope, carbon-12. Radioactive isotopes behave biochemically in the same way as their stable counterparts. For example, radioactive iodine, like the stable form of iodine, is incorporated into the hormone thyroxin, produced by the thyroid gland.

Human-made Sources

   Humans also contribute to environmental radiation through buildings, consumer products, medical and dental treatments, energy production, fallout from nuclear weapons tests, and some industrial processes.

Potential Sources of Radiation Exposure
(mrem per year, unless otherwise indicated)

1. Natural Sources

Source	Variables	Potential Exposure (mrem)	Your Exposure (mrem)
cosmic rays	add 1.3 mrem for every 100 m above sea level	26	_____
rocks and soil	Atlantic and Gulf coasts	16	
	eastern slope of Rockies	63	
	rest of U.S.	30	
			_____
food		39	_____
radon	varies from area to area	200	_____
airplane travel	multiply by # hours of travel	0.5 per hour	_____
subtotal			_____

2. Human Sources

Sources	Variable	Potential Exposure (mrem)	Your Exposure (mrem)
consumer products	tobacco, exposure of lungs for average smoker	1300	_____
	domestic water supplies	1-6	_____
	building materials	3.6	_____
	mining and agricultural products	<1	_____
	natural gas heaters and ranges	0.2	_____

medical X rays	chest	6	_____
	dental	<1	_____
nuclear	living within 80 km (50 mi) of a nuclear facility	.006	_____
	transportation of nuclear materials	20	_____
	fallout from nuclear weapons testing	1	_____
subtotal			_____
totals	subtotal natural sources		_____
	subtotal human sources		_____
	total		_____

Source: National Council on Radiation Protection and Measurements, 1987, *Ionizing Radiation Exposure of the Population of the United States* (Report #93), Bethesda, MD: National Council on Radiation Protection and Measurements.

RESOURCES

Activity #29: Water, Water, Everywhere?

The Earth has such an abundance of water that it is sometimes called the "water planet" and appears blue when seen from space. Approximately two-thirds of the surface of the Earth is covered by oceans, and water is a renewable resource that is recycled continuously by the water cycle. In spite of this, water supply is a problem for many people and is expected to become more so as the population grows. By 2000, global water use is expected to be at least two times as great as it was in the 1980s.

The water supply in any particular part of the world depends upon precipitation, evaporation, runoff, flow from rivers and streams, and underground reservoirs, called aquifers. The ways in which people use water also affects supply. For example, some countries use large amounts of water for irrigating crops or manufacturing goods. When water is removed upstream, those people who are downstream have less water available. Even among highly-developed countries in the northern hemisphere, water use varies considerably. Canada, which uses about 25% less water domestically than the United States, uses about twice as much as Europe.

Where water is scarce, there can be a delicate balance between population and water resources and conflicts can develop among parts of the population competing for the water. In our own country, access to water has been the basis of numerous controversies, especially in the drier, western regions.

☞In this activity, you will model the distribution of the global water supply, compare water use in different countries, and estimate the amount of water you use on a daily basis.

✔MATERIALS: 1 empty plastic liter bottle
                             1 medicine dropper
                             1 glass measuring cup, metric
                             4 empty glass jars, various sizes
                             masking tape
                             marking pen, black

⊘TIME: 1 1/2 hours for activity, 1 hour for followup

❀ACTIVITY A:

1.   Use the masking tape to prepare the following labels, "Total Water in the World," "Oceans + Saltwater Lakes," "Rivers + Freshwater Lakes," "Subsurface Water," "Sea Ice + Glaciers," and "Atmospheric Water Vapor."

2.   Fill the empty liter bottle with clean water. This represents the total amount of water in the world. Label the bottle. Use the table below to determine how much of this water represents the amount found in sea ice and glaciers. Use the measuring cup and/or dropper to remove the appropriate amount and place it in one of the glass containers. (Twenty drops from the dropper equals one milliliter.) Label the container "Sea Ice + Glaciers." Do the same for "Subsurface Water," "Rivers + Freshwater Lakes," and "Atmospheric Water Vapor." The water left in the liter bottle represents the proportion of the world's water found in oceans and saltwater lakes. Remove the label "Total Water in the World" and replace it with "Oceans + Saltwater Lakes."

Distribution of the World's Water

Category	Percent of Total Water Supply
oceans + saltwater lakes	97.20
sea ice + glaciers	2.15
subsurface water	0.63
rivers + freshwater lakes	0.01
atmospheric water vapor	0.001
Total	100.000

Source: *Earth: The Water Planet.* Copyright © 1989,1991 by the National Science Teachers Association. Used by permission.

Water Use per Capita (Selected Countries)

Country	Liters/day
Iraq	12,582
United States	5,947
Spain	3,230
Mexico	2,478
India	1,683
United Kingdom	1,395
Vietnam	224
Nigeria	122

Source: 1992. *Information Please Environmental Almanac.* Copyright © 1991 by World Resources Institute. Reprinted by permission of Houghton Mifflin Company. All rights reserved.

❀ACTIVITY B:

1.   Remove the label from the liter plastic bottle used in Activity A and empty the water. Attach a strip of masking tape to the side of the bottle from the bottom to the top. Place 10 ml of water in the bottle, mark the water level on the tape and label it "Nigeria."

2.   Use the data in the table "Water Use per Capita (Selected Countries)" on page 128 to calculate the number of milliliters needed to represent water use in each of the countries in the table. Add enough water to the 10 ml to represent per capita water use in Vietnam. Mark and label the tape. Continue adding water and marking the level representative of each country until you have done so for all of the countries in the table.

❀ACTIVITY C:

Use the data in the table below to calculate the amount of water you use on a daily basis.

Average Daily per Capita Water Use in the United States

Activity	Liters Used
Bath	130
Bathroom sink (per minute)	7.6
Dishwashing (machine per load)	45
Dishwashing (by hand, water running, per meal)	76
Drinking (per day)	2
Laundry (per load)	150
Shower (per minute)	19
Toilets (per flush)	20
Water running, kitchen sink or garden hose (per minute)	38

Sources: Environmental Protection Agency, 1992, *Fact Sheet: 21 Water Conservation Measures for Everybody*, Washington, DC: U.S. Envrionmental Protection Agency; American Water Resources Association and U. S. Geological Survey, 1993, *Water: The Resource That Gets Used for Everything*. Washington, DC: U. S. Geological Survey.

❧FOLLOWUP:

Answer the questions below.

FOR ACTIVITY A:

1.   Did you have difficulty measuring out the amounts for each of the portions of the total water? Why? What does this tell you about the availability of water for human consumption?

2.   Which of the portions of the total water are available for human use? Is it technologically feasible to make other portions available to humans? Which ones? How?

3.   In a paragraph, summarize what you have learned from this activity about the global water supply.

FOR ACTIVITY B:

1.   How does the water use per capita in Nigeria compare with that in Iraq? the United States?

2.   How can you account for the difference in water use between Iraq and the United States?

3.   In a paragraph, summarize what you have learned from this activity about water use in different countries.

FOR ACTIVITY C:

1.   What is your estimate of your daily water use? How does it compare with the average of 340 liters of water per person per day in the United States? Remember that this represents only direct use; much more is used in manufacturing, agriculture, etc. For example, it takes 380 liters of water to produce one glass of milk and over 1000 liters to produce a Sunday newspaper. To find the total water you use indirectly as well as directly, multiply your daily use by 20.

2.   In a paragraph, explain what you can do to conserve water.

Activity #30: Living Resources

Imagine going to a tropical island paradise and finding no birds. That almost unbelievable situation is what you would find on the island of Guam. Since the end of World War II, all birds have disappeared from the island, victims of a brown tree snake, which was accidentally introduced to Guam from the Solomon Islands. In fact, bird species are disappearing in most areas of the world, although perhaps not as precipitously and dramatically as in Guam.

Because most birds are active during the day, people are more aware of them than they are of other wildlife. Birds are also attractive and have a large following; almost everyone enjoys the sight and sound of birds. Unfortunately, species of other groups of plants and animals, including those less noticeable, less attractive, but no less important, are disappearing as human activities destroy habitats and upset delicate ecosystems.

Among the less conspicuous and less popular animals that are threatened are the amphibians—frogs, toads, and salamanders. Although the numbers of species of amphibians is relatively small, they are numerous and are very important links in food chains. In New York State, the biomass of redbacked salamanders alone is greater than that of the very obvious, white-tailed deer, which are now so numerous as to be a nuisance. Vertebrates are not, however, the only group of organisms with endangered species. Invertebrates, plants, and microorganisms, include threatened species that also have vital roles in maintaining the ecosystem. Many scientists believe that the loss of variety, or **biodiversity**, is one of the most critical environmental issues.

☞In this activity, you will learn about the rates of extinction and study endangered species.

☾TIME: 4 hours for activity, 1 hour for followup

☙ACTIVITY:

1.   On the graph of "The Rising Rate of Bird Extinctions," (see page 133) make a line parallel to the vertical axis. Label the new axis "World

Population (in millions)." Mark equal intervals along the axis, beginning with 550 at the bottom and ending with 6500. Use the data on world population from the table, "World Population: 1600-2000" to plot a graph of the changes in the world population.

2.   Visit a zoo or wild life preserve and select one endangered animal to study. Using resources available at the zoo or a local library, answer the questions on the "Endangered Species Profile" (see page 134).

3.   Find out the names of endangered species in your area and select one. Answer the questions about the species on the "Endangered Species Profile."

World Population: 1600-2000

Year	Population (in millions)
1600	545
1700	610
1750	760
1800	1000
1850	1211
1900	1628
1950	2517
2000	6261

Sources: (for 1600-1700) C. McEvedy, & R. Jones, 1978, *The Atlas of World Population History*, London: Penguin Ltd. Copyright © Colin McEvedy and Richard Jones, 1978. Reproduced by permission of the Curtis Brown Group Ltd, London; (1750-2000) Population Reference Bureau, 1992.

✎FOLLOWUP:

What conclusions can you draw from a comparison of the graphs of the growth of the human population and the rate of bird extinctions? What evidence do you have for relating the two effects? Could the similarities in the two graphs be the result of coincidence?

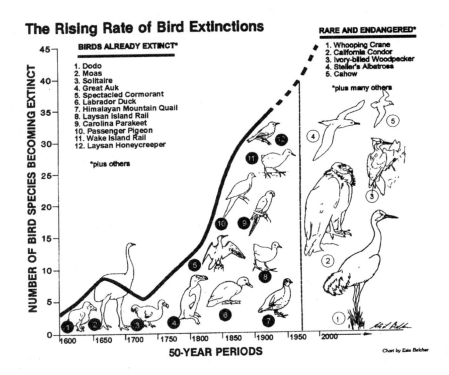

The Rising Rate of Bird Extinctions

RARE AND ENDANGERED*

BIRDS ALREADY EXTINCT*

1. Dodo
2. Moas
3. Solitaire
4. Great Auk
5. Spectacled Cormorant
6. Labrador Duck
7. Himalayan Mountain Quail
8. Laysan Island Rail
9. Carolina Parakeet
10. Passenger Pigeon
11. Wake Island Rail
12. Laysan Honeycreeper

*plus others

1. Whooping Crane
2. California Condor
3. Ivory-billed Woodpecker
4. Steller's Albatross
5. Cahow

*plus many others

NUMBER OF BIRD SPECIES BECOMING EXTINCT

50-YEAR PERIODS

Chart by Este Belcher

Source: G. Nilsson, 1990, *The Endangered Species Handbook*. Washington, DC: Animal Welfare Institute. Reprinted by permission.

Endangered Species Profile

1. Common name _____

2. Scientific name _____

3. Describe the animal's habitat _____
    _____

4. List the animal's prinicipal sources of food _____
    _____

5. List the animal's principal enemies in the wild _____
    _____

6. What elements in the animal's habitat are most important for its
    survival? _____
    _____

7. What factors are threatening or endangering its survival? _____
    _____
    _____

8. What steps are being taken to protect and preserve the animal?
    _____
    _____

9. What are its chances of survival? _____
    _____
    _____

10. What other steps could be taken to preserve this species? _____
    _____
    _____

11. Why is it important to preserve this species? _____
    _____
    _____

Activity #31: Food for Thought

Throughout the world, approximately 700 million people lack the amount and kind of food they need to have a healthy life and carry out their daily activities. The problem is especially acute in developing countries in Africa, the Near East, and Latin America. Whether food supply can keep pace with population growth is controversial. Although many economists are optimistic, most environmentalists are less so. For people in developing areas to maintain their current average diet of 2,500 calories per day while the population doubles, food production will have to increase by slightly more than twice. To bring their calorie intake up to the world average, food production will need to triple. And, if people in developing countries are to have a diet equivalent to those in developed countries, food production will have to increase five times. Even if one or more of these goals are technically feasible, the question remains whether they can be achieved without irreparable damage to the environment.

☞In this activity, you will compare your diet to that of people in other parts of the world.

 You may wish to use data you collected for Activity #22 in this activity.

✔**MATERIALS:**   calorie chart
information on protein content of foods (use the labels on foods you use or consult charts in a diet book)

⊘**TIME:**   1 hour for activity, 1 hour for followup

❁**ACTIVITY:**

1.   Keep track of the food that you eat during one day. If you did Activity #22, you may use that record for this activity. Estimate the number of calories you ate during the day.

2.   Estimate the number of grams of protein that you ate in the day as closely as you can. You will find protein content on food labels and you can

also find information on protein content in books containing calorie charts and other diet information.

3.   Assume that you live in Bangladesh with a daily intake of 1900 calories, 43 grams of protein, and a food supply restricted to grains, such as rice and wheat, and other vegetables. Revise your daily intake as recorded in step #1 to conform to these restrictions also. **Optional**: You may want to eat a diet as restricted as that for people in Bangladesh for one day. Your experience will not be the same as that of someone who is chronically malnourished, but it may give you a better idea of what such a diet is like. **Do not attempt to do this unless you are in good health.** Although a restricted diet for one day is not harmful to most people, it could be dangerous for anyone with special problems.

✎FOLLOWUP:

1.   How does the number of calories you ate compare with the recommended (2300-2900 calories for adult men and 1900-2200 calories for adult women)? to the typical diet of a person in Bangladesh?

2.   Calculate the number of grams of protein recommended for a person of your weight, using the formulas below, and compare it to the number of grams of protein you actually ate:

Your weight (kg) = Your weight (lbs.) ÷ 2.2 lb/kg

Your protein needs (g) = Your weight (kg) x 0.8 g/kg.

Do you eat more or less protein than you need? The average protein consumption in the United States is 111 grams per day; in Bangladesh it is 43 grams per day. How does your intake compare with those values?

3.   The average annual consumption of grains per person is approximately 860 kilograms for an American and 176 kilograms for a person in Bangladesh. The amount of grain required to produce 1 pound of animal food in the United States is: 7 kg for beef, 4 kg for pork, 3 kg for cheese, 2.6 kg for eggs, and 2 kg for poultry and fish. Explain the discrepancy in grain consumption between the United States and Bangladesh. How would you explain the fact that grain consumption in France is almost half that in the United States?

5.   If you attempted to live on a restricted diet for one day, write a brief description of the experience.

Activity #32: Land Use

Providing enough food to feed a growing world population depends on one of our most vital limited resources—land. Although the Earth has a land surface area of 13,128,841,000 hectares (approximately 50 million square miles), only 12% of it is considered suitable for cultivation. The remaining land area is either too wet, dry, cold, or otherwise unsuitable. Scientists estimate that 46-60% of the land suitable for agriculture is currently being used. Opinions differ over how much of the uncultivated lands are marginal (requiring more water and/or energy to be productive than is worthwhile or are ecologically valuable). In addition to these limitations, land is being lost and degraded each year by erosion, unsustainable agricultural methods, logging, deforestation, and development.

Urban areas occupy only about 1% of land area, but their effects through suburban sprawl and pollution are more widespread and growing. The population of urban areas increased by 11% between 1960 and 1990, adding to pressures on nearby land areas. The trend toward urbanization of the human population is likely to have even greater impact on land available for agriculture in the future. The additional area that can be used for agriculture does not seem adequate to feed a population double the present size without an increase in yield per hectare.

How land is used today is crucial to future generations. In the past few decades, the effects of land use changes have become global. As we have attempted to support more people, we have affected rainfall, amount of topsoil and soil quality, air and water quality, tropical rainforests, the ozone layer, numerous species, and possibly global temperatures. Many current environmental conflicts center on land use: from wetlands to landfills, from housing developments to nature preserves, from public grazing lands to strip mines. Although the cumulative effects are global, each change in land use begins with local decisions.

☞In this activity, you will survey land use and make suggestions on future land use decisions for an area near you.

✔**MATERIALS:**   maps of area
                        camera, video camera, cassette recorder (optional)

⊘**TIME:**   8 hours for activity (minimum), 2 hours for followup

❀**ACTIVITY:**

1.   Select a route through an area that passes through several different land use areas. You could select a site that includes a transition between city and suburb, between suburb and country, between residential and industrial areas, or one that includes several transitions. It may be a large area that extends for several miles or a smaller area of a few tenths of a mile. The route you select should follow, as closely as possible, a straight line.

2.   When you have selected your route, you should make a preliminary tour of it to determine how many sampling sites are necessary to obtain a representative sample of the various types of land use. You also must decide what types of information you will need to gather to understand land use issues in the area. It may help you to consider information in the broad categories of physical, social, and environmental setting. (See box on next page for a partial list of suggested topics within these categories.)

3.   Once you have identified the information you want to collect, you should determine how you will collect it. You may use your own observations, interviews with people in the area, photographs, map, audio or video tapes, historical data, or data provided by local governments.

✎**FOLLOWUP:**

1.   Prepare a land use map by tracing a published map of the area and adding outlines of the areas devoted to different uses. As an alternative, you can make a transparency of the published map and create successive overlays on separate transparencies; on each transparency, trace the outline of one land use area. If the map is larger than a piece of legal-size paper, use a copier to reduce it to a manageable size.

2.   Prepare tables and graphs of the data you collected on each land use area and summarize information obtained from interviews, photographs, historical documents, and other sources.

3.   Write a report in which you summarize the current uses within the area you studied and make recommendations for future land use changes.

Information for Land Use Analysis or Planning

I. Physical Setting
    A. topography and landforms
    B. climate
    C. soil characteristics; erosion, stability
    D. vulnerability to natural disasters, such as floods, earthquakes
    E. vegetation
    F. exposure to air and light
    G. water resources

II. Social Setting
    A. social services, e.g., trash and garbage collection, water, sanitation, lighting, snow removal, road repair, fire fighting, schools
    B. zoning regulations/restrictions
    C. transportation, traffic volume and flow; parking facilities
    D. social uses: residential, commercial, industrial, agricultural, educational, recreational
    E. population density
    F. history of area; historical and archeological sites in area
    G. economy of area, opportunities for employment, median income

III. Environmental Setting
    A. natural habitats, wildlife
    B scenic and recreational qualities
    C. evidence of pollution: smog, exhaust odor, noise, chemicals
    D. environmental hazards: disease-carrying pests, such as rodents; high voltage lines or high pressure gas lines; construction sites; abandoned areas with dangerous equipment

POLLUTION

Activity #33: Stream Quality Indicators

Freshwater is only a small part of the Earth's supply of water, yet it is a vitally important resource. Lakes, ponds, rivers, and streams provide habitats for many types of organisms, and they provide drinking water, food recreation, and aesthetic experiences for people. Rivers and streams carry nutrients and minerals vital to life in the oceans, and throughout history they have been important in transportation.

Many human activities are disturbing the delicate ecological balance in freshwaters. They threaten the survival of organisms that live there, as well as human health. Each freshwater species is adapted to a range of conditions—temperature, dissolved oxygen, and pH being the most important. While some organisms have a wide tolerance, many can survive only within a narrow range of environmental conditions. Key organisms, adapted to different conditions, can be used as indicators of water quality. When organisms that require a pH range of 7 to 8 are not found, it may indicate that the water is more acid or more basic. Carp are able to live in water that is too warm and too turbid for many other fish species, whereas trout prefer cold, clear water.

About two-thirds of water pollution results from agriculture, followed by human and industrial wastes. Removal of trees, by deforestation or fire, increases runoff and brings more organic materials to the waters. When a body of water becomes overloaded with organic matter, the oxygen level of the water decreases. Fertilizers and other materials rich in phosphates can result in excessive algae growth and low oxygen levels. As the oxygen becomes scarce, organisms die, adding a further burden of organic material to the water.

☞In this activity you will assess the water quality of a stream and its suitability as a habitat for organisms.

✔**MATERIALS:**   magnifying lens
plankton net (see below)
dip net (see below)
thermometer
orange
heavy string

shallow white or clear plastic containers
2 identical small glass jars with screw-type lids
      (fill one with clean tap water before you
      go)
1 large glass jar with screw-type lid
measuring tape or meter stick
2 watches with second hands
an unlined index card on which you have drawn
      several thin black lines at half-
      centimeter intervals
pH paper or indicator solution (see notes on
      indicator solution in Appendix at end of
      book)
field manuals for freshwater organisms
      (optional)

⊘TIME: 4 hours for activity, 2 hours for followup

❀ACTIVITY:

1.   Select a stream in your area and prepare your equipment. If you do
not have the equipment listed, you can improvise as described below:

> Plankton net: Cut a piece from the foot end of a nylon stocking and
> tie or sew it to a circular metal rim made from a coat hanger or
> piece of wire. Slip a small bottle into the toe of the stocking and
> secure the stocking around the bottle with a rubber band. Tie 4
> pieces of heavy string about 6 inches long to the metal rim. Tie a
> loop at one end of another piece of string, about 3 feet long, and tie
> each of the short pieces to the loop. If necessary, you can dispense
> with the rim and string and drag the piece of stocking itself.

> Dip net: You can also use a piece of nylon stocking tied to a
> circular metal rim to make a dip net. However, instead of tying
> string to the rim, extend both ends of the wire forming the rim so
> that they protrude at right angles to the rim. Twist these around a
> long piece of wood, preferably a dowel or mop handle. You can
> also use a kitchen sieve with a fine mesh as a dip net, although the
> handle will be short.

2.   To estimate the velocity of the stream, you will probably need to
have someone help you. Mark two places on the bank, 30 meters apart. If

it is not feasible to mark off 30 meters, mark the greatest distance you can. Measure the distance and record it in your notebook. One person should then stand at each mark. The one who is upstream should drop the orange into the water, as close to the middle of the stream as possible, and make note of the time. The downstream person should make note of the time at which the orange passes the second mark. Divide the distance, in meters, by the time, in seconds, to find the velocity of the stream.

3.   Take the temperature of the water at several places along the bank and at different depths. Allow at least 1 minute each time for the thermometer to adjust to a new location. Find the average temperature.

4.   Collect some water in the empty small glass jar. Place the card you have prepared behind it and then behind the jar of tap water. Compare the clarity of the samples. Use your observations on temperature, velocity, and clarity to estimate the oxygen content of the water. Rate the stream according to the following guide:

Qualitative Estimate of Oxygen Content

Temperature	Velocity	Clarity	Odor	Oxygen Content
cold (below 13°C)	fast	clear	none	high
warmer (13-20°C)	moderate	slightly cloudy	slight	moderate
warm (above 20°C)	not moving	cloudy	strong	low

5.   If you have pH paper, measure the acidity of the water. Dip the end of a strip of paper in the water and compare with the color scale on the container of pH paper. If the pH is below 7, the water is acid; if above 7, it is basic; and, if close to 7, it is neutral.

6.   Fill the large jar half full with water from the stream. Drag the net through the water and add water from the plankton net jar and debris and organisms from the dip net to the water in the jar. Add some plant material from the stream. Try to collect at various types of locations in the stream, especially where the water flows over stones. Close the jar, making sure to leave plenty of air space above the water level. You can examine the organisms you collect at the stream or wait until you have returned to home or school.

7.   Turn over rocks, leaves, and sticks at the edge of the stream and collect any organisms you find in the small glass jars. (Be sure to empty the tap water from the one before you add stream organisms.) Put some stream water into each jar. If you will not be able to identify the specimens you collected for many hours or days, you can put them in jars with rubbing alcohol. However, this may be hard on the soft-bodied organisms and make them more difficult to identify.

8.   When you are ready to identify your specimens, spread them on the flat containers. If your containers are clear, you may find you can see the organisms better if you put a piece of white paper under the container. Compare the organisms you can see with your unaided eye and with the magnifier to those shown in "Indicator Stream Organisms" (see page 148).

      If you have a microscope, look for microscopic plants and animals.

✎FOLLOWUP:

1.   Use "Indicator Stream Organisms" to identify the organisms you have found. Classify them into the three groups on the chart. Count the number of organisms in each group. List all organisms you cannot identify in a fourth group. You may be able to identify those from the field manuals, or you can draw a picture of them, classify them by type, or give them your own names. The common types of organisms are:

Characteristics	Type of Organism
three pairs of legs, body in 3 main sections, 1 pair of antennae, usually 2 pairs of wings	insects
hard outer shell, soft body inside	molluscs
usually 2 body sections, 2 pairs of antennae	crustacea
soft-bodied, many segments	segmented worms (Do not confuse worms with insect larvae, which may be long and worm-like. Insects will have jointed legs.)

2.   Complete the "Stream Quality Form" and rate the quality of the stream.

Stream Quality Form

Group	Number of Types of Organisms in Group (N)	Group Index Value (I)	Assessment Value (V = N x I)
1		3	
2		2	
3		1	
Cumulative Assessment Value (V1 + V2 + V3)			

Stream Quality Rating

Cumulative Value	Stream Quality
>22	Excellent
17-22	Good
11-16	Fair
<10	Poor

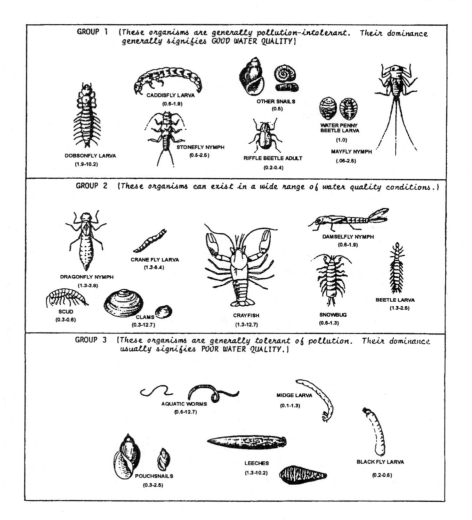

Indicator Stream Organisms
(Numbers in parentheses are size ranges in centimeters.)

Source: Courtesy of Ohio Scenic Rivers Program, 1993, *A Guide to Volunteer Stream Quality Monitoring*, Columbus, OH: Ohio Department of Natural Resources.

Activity #34: Lichen Indicators

Lichens are associations of algae and fungi whose lives are so closely entwined that they appear to be single organisms. This symbiotic relationship is generally considered to be mutually supportive—with the algae providing food through photosynthesis and the fungi providing moisture, support, and minerals. However, the fact that the fungi cannot survive by themselves, while the algae can, raises the question of whether this is a truly mutual arrangement.

Lichens are a living paradox. Although found in every habitat, including those with the harshest, most extreme conditions, lichens are extremely sensitive to certain environmental factors. In one experiment, more than half of the lichens transplanted from the bases of trees to a height of 4 feet were adversely affected, probably a result of less stable temperature, moisture, light, and wind conditions. Lichens are also sensitive to air pollution, particularly the acid pollution associated with high levels of sulfur dioxide. Their absence from some urban areas is so complete that those areas have been called "lichen deserts." Because of their sensitivity to pollution, lichens are good indicators of air quality.

Lichens occur on bare rock surfaces, tree trunks, concrete structures. Although a few lichens with very distinct growth forms, such as the "British soldiers," are easily recognized, identification of most of the 20,000 known species of lichens is extremely difficult. For nontechnical purposes, lichens are grouped into three main types, based on the form in which they grow. **Fruticose** lichens are upright or hanging forms, often shrubby or stringy looking, with no distinct upper and lower surfaces. **Foliose** forms are flat, leaf-like and have distinct upper and lower surfaces. **Crustose** lichens form tightly-adhering crusts on the surfaces on which they grow. Even if a lichen appears leaf-like, as long as it cannot be separated from its surface, it is a crustose lichen. Crustose lichens are more tolerant of pollution than are the other two.

☞ In this activity you will conduct a lichen survey of two or more areas and use the types and size of lichens present to estimate air quality.

crusty

shrubby   leaflike

Types of Lichens

Source: Reprinted, with permission, from the Air & Waste Management Association's *Environmental Resource Guide-Air Quality, Grades 6-8*, Air & Waste Management Association, One Gateway Center, Pittsburgh, PA 15222, 1991.

✔**MATERIALS:**   transparency sheet
                       washable marker
                       damp sponge
                       calculator (optional)

☺**TIME:**   2 1/2 hours for activity, 1/2 hour for followup

✻**ACTIVITY:**

1.   Use the grid sheet on page 153 to make a transparency. (You can usually have this done at commercial copier stores.) Prepare a data table with 5 columns (no lichens, grey-green crusty lichens, orange crusty lichens, leafy lichens, and shrubby lichens) and one row for each study area.

2.   Select two or more areas in which to conduct your lichen survey. The areas should be ones that you suspect differ substantially in air quality.

3.   In each study area, locate lichens one at a time. Classify the lichens according to the five groups in the figure, "Indicator Lichens" (see page 152) and tally the numbers in each group.

4.   Hold the transparency up to each lichen without touching the lichen. Use the washable marker to trace the outline of the lichen. Count the number of full squares within the outline. Add the total area of partially covered squares by estimating the proportion of each that is covered to the

nearest tenth of a square. Record the total area for the lichen. Remove the marker from the transparency and proceed to the next lichen. You should measure at least 10 lichens in each study area.

5.   Find the average area of lichens in each study area.

✎FOLLOWUP:

1.   Compare the data in the table you made of types of lichens to the figure, "Indicator Lichens" to estimate the air quality in each area you studied.

2.   Compare the average area of lichens for each study area with the ratings below and estimate the air quality in each area you studied.

Average Area (cm^2)	Air Quality
10-12	excellent
7-9	good
4-6	fair
0-3	poor

3.   Do the two methods for estimating air quality give you the same results? If not, how different are they? How can you explain the differences?

4.   How do the different study sites compare in air quality? What factors could account for any differences you found?

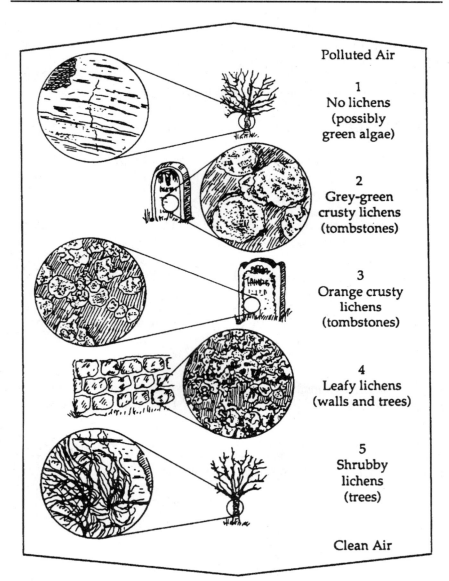

Indicator Lichens

Source: Reprinted, with permission, from the Air & Waste Management Association's *Environmental Resource Guide-Air Quality, Grades 6-8*, Air & Waste Management Association, One Gateway Center, Pittsburgh, PA 15222, 1991.

Grid Sheet for Transparency

Activity #35: Chemical Pollutants

Close to 100,000 chemicals are in use, with hundreds of new ones added each year, making this as much a chemical society as an information society. As with most technological advances, chemicals have brought great benefits to human life, and have created new problems as well. Chemical technology has provided people with antibiotics, miracle materials, better preserved food, pesticides, and fertilizers. But it has also adversely affected human health and the environment. Birth defects, cancer, mental retardation, and diseases of the nervous, endocrine, and immune systems have been linked to chemicals in the environment.

Among the many chemicals in use today are natural ones that are being released at unnatural rates, as well as synthetic ones. Lead, for one, is being released into the environment at 300 times the natural rate. While the effects of some of the chemicals are well-known and those of others are suspected, there is no information on three-fourths of the ones now in use.

Dramatic effects of chemical pollution, such as the link between mercury and severe neurological disturbances, are only the tip of the iceberg. Scientists now recognize that many chemicals affect living things in subtle ways that are not easily documented. Furthermore, the chemicals can react with each other to produce new dangers. Chlorine introduced into municipal water systems can react with organic chemicals from industrial sources to produce carcinogenic substances. Chemicals that are fat-soluble tend to accumulate in the fatty tissue of organisms and then be passed on to organisms that feed on them. At each step in the food chain, the chemicals become more concentrated, a phenomenon known as **biomagnification**.

Because chemicals can be dangerous to human life and it may take years, or even decades, to determine their effects, scientists test chemicals on other organisms, especially those, such as bacteria, fruit flies, and mice, that reproduce rapidly. Fast growing plant tissues, such as young seedlings and roots, are also used for evaluating the biological effects of chemicals.

☞In this activity, you will use the growth of roots on onion bulbs to test for environmental pollutants.

✔**MATERIALS:**   10 (or more) small white onions
                            glass jar or plastic cup (1 per onion)
                            samples of water from various sources
                            10 jars with screw type lids
                            labels and marker
                            toothpicks
                            single-edge razor blade
                            ruler

☾**TIME:**   3 hours for activity, 1 hour for followup

❀**ACTIVITY:**

1.   Peel off the loose, outer layers of each onion. Locate the root end and cut a <u>thin</u> (approximately 1 mm) slice from the root end with the razor. **Do not cut off the bases of the roots.** After you have made your slice, you should still see the bases of the roots as circular or eliptical shapes, as shown in the diagram below. You may find it easier to make your slices of consistent thickness if you use something 1 millimeter thick to guide your razor cut. (The type of cardboard used at the base of a pad of paper is a good choice.)

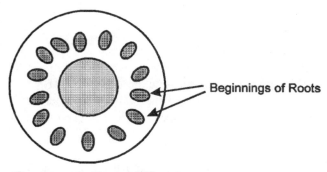

**Cross-section of Root
End of Onion Bulb**

2.   Place each onion on top of one of the jars. If the onions are too small to rest on the rims of the jars, stick toothpicks at right angles to the surface midway between the top and bottom of the onions and rest the toothpicks on the rims.

3.   Collect water samples from various sources, such as streams, faucets, swimming pools, etc., including boiled tap water. Label the jars and

fill each jar with one of the water samples so that the base of each onion is in the solution. Be sure to use the boiled tap water in one of the jars. (Boiling drives off any carbon dioxide in the water that could make the water acidic and should result in water with a pH close to 7.) Cover each jar with a lid so that the water will not evaporate during the course of the experiment.

4.   Put the jars in a place with little or no light. Empty each jar every day and replace with fresh water of the same type.

5.   In about 5 days, when new roots have begun to grow, measure the length, in millimeters, of the roots of each onion. Record your measurements in a data table. Measure and record the length of the roots every day for the next 10 days.

If you cannot locate enough different water samples, you can try one of the following methods: (1) make different dilutions of one sample and test each one (be sure to make enough of each dilution to be able to change the water every day), or (2) use common household substances, such as salt or vinegar, to make a range of dilutions to test. You can measure the composition of each dilution as a percentage by weight or volume. For example, a 1% salt solution would have 1 gram of salt in 100 g of water; a 1% vinegar solution would have 1 ml of vinegar in 100 ml of liquid. If you do not have a scale that can measure in grams, you can obtain a mass of approximately 1 gram of salt by the method described on pages 112 and 113.

You can relate this activity to Activity #5: The Power of Doubling, by graphing average root length for days 1-10, with root length on the vertical axis and days on the horizontal axis.

You can do this activity in conjunction with Activity #33 by using water from the stream as one of the samples.

✎FOLLOWUP:

1.   Make a graph of your results. If you have used different types of water in each jar, make a bar graph with the type of water on the horizontal axis and the length at 10 days on the vertical axis. If you have used different dilutions, make a line graph with the strength of solution on the horizontal axis and the length at 10 days on the vertical axis.

2.   Write a brief report of your results and discuss explanations for the results. Why was it important to use boiled tap water in one jar? Include a description of possible errors in the experiment and how the experiment could be improved.

An Example of Biomagnification

Organism	PCB Level (ppm)*
plant plankton	0.0025
zooplankton	0.123
rainbow smelt	1.04
lake trout	4.83
herring gull eggs	124

Source: L. Botts, & B. Krushelnicki, 1987, *The Great Lakes: An Environmental Atlas and Resource Book.*, Toronto: Environment Canada; Chicago: U.S. Environmental Protection Agency.

Activity #36: Greenhouse Globe

The temperature of the Earth is affected by its distance from the sun and the presence of its atmosphere. Without an atmosphere, the planet would have more extreme temperatures, such as those found on the moon. While the side of the moon facing the sun has temperatures of 127°C (260°F), the side away from the sun has temperatures of -173°C (280°F) Without an atmosphere, the average surface temperature of Earth, 14°C (57°F), would be 33°C (59°F) colder than it is.

The energy from the sun that reaches Earth is primarily in the form of rays of visible and ultraviolet light. When these encounter molecules in the atmosphere and on the Earth's surface, they generate heat, or infrared rays. Some gases in the atmosphere allow visible and ultraviolet light to pass through, but trap heat energy. Known as "the greenhouse effect," this heat-trapping phenomenon helps to warm Earth to a comfortable temperature.

The major contributor to the greenhouse effect is carbon dioxide. Although its concentration in air is very low—0.035% or 350 parts per million (ppm)—it has increased from 315 ppm since 1958. Scientists estimate that carbon dioxide levels were about 275 ppm in the early part of the 18th century. The increase is attributed to human activities, such as combustion of fossil fuels and deforestation. Many scientists are concerned that continued increases in atmospheric levels of carbon dioxide will result in a global warming of 1.5°C - 4.5°C (3°F - 9°F) by the middle of the next century. A change in average global temperature of a few degrees could have a substantial effect on climate, the biosphere, and human life. Furthermore, past climate changes of this magnitude have occurred over thousands of years, rather than decades.

☞ In this activity, you will study some sources of carbon dioxide.

✔**MATERIALS:**   16 3-4 oz. clear plastic cups
6 sheets of unlined white paper
device for measuring small quantities of liquid
              (a medicine dropper or small measuring
              spoon)
metric measuring cup

indicator solution (bromothymol blue solution or
you can make your own by following
the directions in Appendix
at the end of the book)
test substances: (approximately 20 ml each)
chlorine bleach, ammonia, milk of
magnesia, borax, baking soda, distilled
or boiled water, coffee, orange juice or
fresh tomatoes, vinegar, lemon juice
1 short, sturdy candle, able to fit inside jar
1 8-oz. jar with screw-type lid
straw
1 sheet of heavy duty paper or 1 manila folder
1 balloon (the sturdiest one you can find)
masking tape
automobile

⊘**TIME:**   2 hours for activity, 1 hour for followup

❀**ACTIVITY:**

1.   Detecting carbon dioxide and pH: Carbon dioxide combines
with water to produce carbonic acid. Therefore, it can be detected in
liquids with an acid-base indicator. The strength of an acid or base is
measured on the pH scale, which ranges from 0 to 14. Substances with a
pH greater than 7 are basic; those with a pH below 7 are acidic.
Substances that are neither acidic or basic, i.e., neutral, have a pH of 7.
Before you use the indicator to test for carbon dioxide, you will observe
the effects of a number of common substances of different pH values on
the indicator.

For each of the test substances, place 30 ml of water in a plastic cup.
Label each cup with the name of the test substance. Place the cups on
plain white paper. Add measured small amounts of the indicator to one
of the cups until the liquid has a faint, but distinct, color. Add exactly the
same amount of indicator to each cup. Note the resulting color. Now add
a measured amount of lemon juice to one cup. With a clean measurer,
add the same amount of each test substance to each of the other cups.
Record the colors in the cups in the table on page 161. (You will need to
dissolve the borax and baking soda in small amount of water before
testing their pH.)

Indicator Table

Substance	Approximate pH	Color
chlorine bleach	12-13	
ammonia	11	
milk of magnesia	10	
borax	9	
baking soda	8	
distilled or boiled water	7	
coffee	5	
orange juice or tomatoes	4	
vinegar	3	
lemon juice	2	

2.   Burning: Burning requires an organic fuel (a carbon-containing compound) and oxygen.

a.   Put a candle in a jar and light the candle. Allow it to burn for a few minutes. Then cover the jar. When the candle goes out, observe the inside walls of the jar. Record your observations.

b.   Prepare two cups of water and indicator, using the same measurements as before. Pour the colored liquid from one cup into the glass jar with the candle. Light the candle and allow it to burn uncovered for 2 minutes. Slide the cover over the top so that it almost closes off the candle's air supply, and allow the candle to burn for 2 more minutes. Cover the jar and screw the cap on tightly. Shake the jar so that the liquid can mix with the air in the jar. (Don't worry about the candle. It will just bounce around inside.) Unscrew the lid and pour the liquid back into the small cup. With the two cups on a piece of white paper, record the color of the liquids.

3.   Respiration: Respiration also requires a fuel and oxygen; the usual fuel in living organisms is sugar (glucose). But respiration occurs at lower temperatures and more gradually than burning. Decay is similar to respiration.

> a.   Breathe onto a mirror and record your observations.
> b.   Prepare two cups of water with indicator. Bubble your breath into one cup by blowing through the straw. After a few minutes, put the two cups on a piece of white paper. Record the colors of the liquids.

4.   Combustion of Fossil Fuel

Prepare two cups of water with indicator. Make a cone from heavy duty paper or a manila folder so that one end of the cone just fits over the tailpipe of the car and the mouth of the balloon fits over the other end. Secure the cone with tape. Fit the mouth of the balloon over the smaller end of the cone. Start the car and as it idles, place the large end of the cone over the tailpipe. **You need to do this cautiously. Do not stand right behind the tail pipe. Avoid breathing the exhaust fumes.** When the balloon is partly inflated, remove the cone and pinch the mouth of the balloon to prevent gas from escaping. (You can also twist the neck of the balloon and/or tie it with a twist-tie.) Put the end of the balloon into the liquid in one of the cups and allow the gas to escape slowly. Put both cups on a piece of white paper and record the colors of the liquids.

✎FOLLOWUP:

1.   What two substances are produced by burning? by respiration? How do you know?

2.   Explain how deforestation can affect the amount of carbon dioxide in the atmosphere. Is the effect different if trees are cut and burned or if they are left to rot?

3.   Combustion of 1 gallon of gasoline produces 24 lbs. of carbon dioxide. Based on the number of miles you drive in one year and the number of miles per gallon you get, calculate the number of pounds of carbon dioxide your car produces each year.

4.   The annual production of carbon dioxide worldwide is 20 billion tons. What is the per capita production of carbon dioxide? The United States produces 27% of the carbon dioxide. How many tons of carbon dioxide are produced by the United States alone? What is the per capita production in the United States?

Greenhouse Gases			
Gas	Sources	Lifespan	Current Contribution to Global Warming
carbon dioxide	fossil fuels, deforestation, soil destruction	500 years	49%
methane	cattle, biomass, rice paddies, gas leaks, mining, termites	7-10 years	18%
nitrous oxide	fossil fuels, soil cultivation, deforestation	140-190 years	6%
chlorofluoro-carbons	refrigeration, air conditioning, aerosols, foam blowing, solvents	65-110 years	15%
ozone	action of sunlight on atmosphere	hours to days	12%

Source: From *The Greenhouse Trap* by World Resources Institute. Copyright © 1990 by World Resources Institute. Reprinted by permission of Beacon Press.

Activity #37: Kirtland's Warbler

The extent of potential levels of global warming on climate are being predicted by computer models, each of which is based on a number of assumptions. Although the models do not always agree, they provide valuable information on the possible consequences of global climate change. Projections or scenarios based on these models are being studied by scientists, even as the models themselves are being refined. The models predict that global warming will cause many communities of plants and animals to change and some species, particularly those already threatened, to become extinct. In fact, some scientists believe that especially vulnerable species may already be showing the effects of global warming and acting as early warning signals of future problems.

☞In this activity, you will investigate the effects that global warming could have on one species of bird.

✔**MATERIALS:**

☺**TIME:**   1/2 hour for activity, 1 hour for followup

✿**ACTIVITY:**

1.   Read the background material about Kirtland's Warbler in the box on the next page.

2.   Answer the questions in the Followup section.

✎**FOLLOWUP:**

1.   It is predicted that a one degree (Celsius) rise in temperature will cause plant species to migrate approximately 100 to 150 kilometers (60-90 miles) to the north. Many scientists predict that global warming will result in a 4° C increase in average global temperatures by the year 2030. How many kilometers will the jack pines need to move to stay within their temperature tolerance?

2.   The radius of the area inhabited by Kirtland's warbler is about 100 kilometers (60 miles). How would a 4° C increase in temperature affect Kirtland's warbler?

Kirtland's Warbler

Kirtland's warbler is named for the man who discovered the first specimen on his Michigan farm in 1851. Like all warblers, it is a tiny bird; an adult male weighs less than 14 grams (about half an ounce). Possibly because it is rare, or because of its beautiful plumage and loud, melodic song, it has been the "pet" bird of Michigan.

The birds winter in the Bahama Islands and migrate north to central Michigan each spring. There the pairs build nests in stands of young jack pines, where they grow on an unusual coarse sandy soil, called Grayling sand, found only in parts of Michigan. The warblers lay three to five tiny eggs, each 18 millimeters (about 3/4 of an inch) in diameter. Often cowbirds lay their eggs in the warblers' nests. When they do, the warbler eggs are literally left out in the cold as their parents incubate the cowbird eggs instead of them. Few young warblers survive this competition.

Jack pines depend on periodic fires for their seeds to germinate, but with the settling of the region by farmers, fire control measures resulted in a decline in the area of jack pine forest. Furthermore, jack pines are extremely sensitive to temperature. Those in the area where Kirtland's warbler lives are at the southern extreme of their range. Most of them are found in the cooler climates to the north.

Scientists do not understand why the warblers require jack pines of a certain age (5-6 years old) and Grayling soil, but the result is a species with very limited ability to survive environmental changes. Since the early 1970s, federal, state, and wildlife agencies have set fires, planted trees, and trapped cowbirds in an attempt to save the warbler from extinction. In spite of these efforts, the numbers have declined. In 1951, there were 432 males, while in 1989 there were only 212.

3.   Forests are known to have adapted to temperature changes in the past of one degree Celsius over 1000 years. Even the most conservative estimates of global warming predict a one degree increase by the middle of the next century. How do you think the temperature increase due to global warming will affect the jack pines, and why?

4.   If the warmer climate causes more frequent forest fires, how might that affect your answer to #3?

5.   Some scientists have compared Kirtland's warbler to a "miner's canary." What is meant by this?

Activity #38: Arctic Warming

The area of land in the far northern latitudes, which includes tundra, bogs, and forests, is only 14% of the total world area, but contains 25% of the carbon stored in the world's soils. The tundra areas themselves, representing only 6% of the world's land area, contain 12% of the global soil carbon. Because the Arctic plays such a disproportionate role in the global carbon cycle, understanding the effects of temperature on the region is critical to predicting the effects of global warming.

In the tundra, all but the top 30-40 centimeters of soil remain frozen all year; the permanently frozen portion is known as permafrost. When the top layer melts in the summer, water cannot drain away readily because of the frozen underlayer, and soils are fairly well waterlogged. Small, low-growing plants adapted to the cold, short growing season, and saturated soils, store carbon through photosynthesis. During the rest of the year, the plants are relatively inactive and they produce little carbon dioxide through respiration. Furthermore, when plants die, the cold temperatures and waterlogged soils inhibit the bacteria responsible for decay, so a great deal of carbon remains stored in the soil. Under conditions of low rates of decay, nitrates and ammonium, which stimulate plant growth, are not released and remain unavailable to plants.

Until recently, therefore, the Arctic returned less carbon to the atmosphere than was removed through photosynthesis; the region was a sink for carbon dioxide (see Activity #18). Recent measurements indicate that the Arctic, particularly the tundra, is now releasing more carbon dioxide than it is storing; that is, it is acting as a carbon dioxide source. Under laboratory conditions, higher levels of carbon dioxide stimulate photosynthesis in some plants. One hypothesis is that the increased release of carbon dioxide is related to temperature increases of 2°C - 4°C observed in northern Alaska and Canada over the last few decades. Scientists from the United States and the former Soviet Union are actively studying the arctic ecosystem to learn more about the potential effects of continued warming.

☞In this activity, you will analyze the effects of global warming on the Arctic and identify possible positive and negative feedback paths.

☻TIME: 1 hour for activity, 1 hour for followup

❀ACTIVITY:

1.   Develop an "effects wheel" for warming of the Arctic. An effects wheel begins with a single term in the center surrounded by terms describing the direct effects of the term in the center. The direct effects may produce other, secondary, effects. Begin by putting the term "warming" in the middle of a blank page. Think of the direct effects of warming on the Arctic and write them around "warming." Connect each direct effect to "warming" with straight lines. Think of the effects of each direct effect and add these to your diagram around the direct effects. Continue, until you can think of no more effects. Your completed diagram should look something like this:

2.   Ecosystems are not static, and changes in the Arctic are bound to occur if the temperature continues to rise. Different responses, however, occur over different time scales; some adjustments are very rapid, while others may require centuries. Using the table on the next page and the effects wheel you created, predict the response of the arctic ecosystem over the next few years, the next few decades, and the next few centuries, assuming that global warming continues for the next 100 years.

Level of Response	Time Scale for Response
bacterial	minutes to days
growth of individual plants	days to weeks
ecosystem: carbon dioxide flux	weeks to years
ecosystem: changes in species composition due to competition	years to decades
ecosystem: migration and evolution	decades to centuries

3.   Some plants respond initially to increases in carbon dioxide with an increased rate of photosynthesis. After a time, however, the accumulated carbohydrates produced by photosynthesis reduce the rate of photosynthesis to the previous level. In addition, fast growth may deplete nutrients in the soil. If this were the case for a large number of arctic plants, how would that affect your answer to question #2?

4. Identify one example each of potential positive feedback and negative feedback in the arctic response to global warming. (You may want to review the discussion of positive and negative feedback in Activity #2.)

✎FOLLOWUP:

1.   Is the warming trend observed in the Arctic in recent decades proof that global warming is occurring as a result of increases in carbon dioxide level in the atmosphere? What other possible explanations are there?

2.   At times in the past, the Arctic has been much warmer than it is now. If that is the case, should we be concerned about the present warming trend? Why, or why not?

Activity #39: Acid Rain

Rainwater is normally slightly acid because it absorbs carbon dioxide as it passes through the atmosphere, forming a weak acid known as carbonic acid. Other gases in the atmosphere, such as sulfates and nitrates, contribute to the normal acidity of rain when they too combine with water. Normal rainwater, with a pH of 5.6 to 6.0, dissolves minerals in soil and rocks, making them available to plants. Combustion of fossil fuels produces additional amounts of carbon dioxide, sulfates, and nitrates. When the resulting acids are added to the natural acidity, the pH of rainwater may go as low as that of pure vinegar, approximately 3.0.

Acid rain (with a pH below 5.6) has an adverse effect on many forms of plant and animal life. It damages leaves, affects growth of roots, and inhibits germination in plants. Many lakes in the northeastern portion of the United States and southeastern Canada are so acid that they can no longer support the normal community of microorganisms, plants, and animals. Lakes with a high content of carbonate in the rocks and water can resist changes in pH and are better able to withstand acid rain, while others have little buffering capacity. Of the 2,800 lakes and ponds in the Adirondack region of New York State, 25% are too acid to support life or are in danger of becoming so. Highly acid water also increases the concentrations of heavy metals, such as lead and mercury, in water, posing a threat to human health as well as to that of plants and animals.

Coal with a high sulfur content is largely responsible for the production of sulfates in polluted air, while automobiles, power plants, and industries relying on petroleum are the main sources of nitrates. Cleaning coal before burning and removing pollutants as they are exhausted through smokestacks have brought sulfate emissions down by 30% since 1980. Reducing nitrate emissions is more difficult because of the widespread use of automobiles and petroleum fuel, but more stringent pollution control regulations have resulted in a 6% decrease since 1978. In spite of decreasing emissions, however, the acidity and nitrate concentration of lakes and streams has increased and the buffering capacity decreased.

☞ In this activity, you will study the effect of acid on seed germination.

✔MATERIALS: 100 seeds of one kind (bean, pea, corn, etc.)
6 plastic dishes or cups
paper toweling
white vinegar
boiled tap water
clear plastic wrap
masking tape (for labeling)
metric measuring cup
metric ruler
graph paper
1 piece of chalk
1 antacid tablet
rainwater, pondwater, or snow samples
(optional)
pH paper, bromothymol blue, or other
indicator solution (optional)

⊘TIME: 2 hours for activity, 1 hour for followup

  If you have not yet done Activity #36, you should do step #1 of that activity before doing this one.

✿ACTIVITY:

1.   Boil 2 liters of tap water for 5 minutes. Set aside 200 ml of the boiled water for a control solution. Use the rest of the water to prepare 200 ml of each of the following test solutions: 50%, 25%, 10%, 5%, and 1% vinegar. (A 5% solution would contain 5 ml of vinegar for each 100 ml of solution. Therefore, to 10 ml of vinegar, add enough water to make 200 ml of solution.) Label each container.

2.   If you have pH paper, measure the pH of each solution, including the control and rainwater or pondwater samples. If you do not have pH paper, use an indicator solution, and determine an approximate pH for each test solution and sample. You may not be able to distinguish differences in pH for the more acid solutions with an indicator, so pH paper is preferable. If you have neither pH paper nor an indicator, make your own indicator solution; directions for this are in the Appendix.

3.   Cut 4 pieces of paper toweling for each dish or cup. Each piece should just fit into the bottom of the dish or cup. Put two pieces of toweling on the bottom of each container.

4.   Measure the length of 10 seeds and record the data. Place the seeds on the toweling. Add enough of the test solution to wet the toweling and to leave a small amount of liquid in the bottom of the container. Cover the seeds with two more pieces of toweling and press the toweling down so that it becomes saturated with the solution. If some of the toweling is still dry, add more of the test solution. Put a piece of clear plastic over the container to prevent evaporation. Repeat for each test solution, control, and sample. Put the containers in a dimly lit area.

5.   Each day, measure the length of the seeds, including any material that has sprouted. Record the number of sprouted seeds. Add more test solution, if necessary, to keep the seeds moist. Continue for 7-10 days, depending upon the type of seeds.

6.   a.   Put a piece of chalk in a cup of vinegar and observe the results. (Chalk is made of calcium carbonate.)
        b.   If you have pH paper, compare the pH of vinegar with that of vinegar to which you have added a crushed antacid tablet. If you are using indicator solution, prepare two cups of indicator. Put the crushed antacid tablet in one cup. Add the same amount of vinegar to each as you did in step #2. Record the colors of the two solutions. Continue adding measured amounts of vinegar to the cup with the antacid tablet. How much vinegar do you need to add to obtain the same color as in the other cup?

7.   If you have collected snow samples, remove the top 1 cm of snow and place it in a clean, covered container indoors. When it has melted, measure its pH with pH paper or estimate the acidity with indicator solution. You can take samples from different areas and compare the results.

✎FOLLOWUP:

1.   Find the average seed length and the percentage of germinated seeds for each day for each test solution, control, and sample. Summarize the data in a table.

2.   Draw a line graph showing change in seed length, with average seed length, in millimeters, on the vertical axis, and number of days on the horizontal axis. You will need to use a different line for each test solution and sample. Draw a bar graph of germination success, with the percentage of germinated seeds on the vertical axis and the concentration of vinegar (or pH) on the horizontal axis.

3.   What conclusions can you draw from your data? What is the optimum acidity for the seeds you used? How do your results compare with those of others on the same type of seeds? other types?

4.   Explain why acid rain is attacking buildings, monuments, and statues. Explain how carbonate rocks can buffer lakes against acid precipitation.

5.   If you tested rainwater or pondwater samples, what can you conclude about the acid rain situation in your area? If you tested snow samples, explain the phenomenon of "snow shock" during the spring when the snow melts. What different effects might a sudden warm spring have compared to a slow melt?

 Instead of using seed germination to test the effects of acid, you can use growth of seedlings (Activity #4) or growth of onion roots (Activity #35).

Effects of Acidity on Freshwater Life	
pH	Effects
7.0	typical pH, normal communities
6.0	trout, clam, crustacea begin to decline
5.5	rainbow and brown trout begin to die; brook trout fail to reproduce; clams and crustacea no longer likely to occur
5.0	interferes with reproduction of most species of fish
4.2	common toad dies
3.5	almost all fish, clams, snails, and frogs unable to survive
2.5	only a few species of midges, fungi, and bacteria can survive
0.0	no life can survive

Source: From page 100 from *Design for a Livable Planet* by Jon Naar. Copyright © 1990 by Jon Naar. Reprinted by permission of HarperCollins Publishers, Inc.

APPLICATIONS

Activity #40: House Trees

Trees are often planted in residential yards to improve the aesthetic environment and to provide habitats for birds and other desirable wildlife. When planted in the proper orientation to the house and the sun, trees can also help to moderate the climate in and around the house. Shade trees that block sunlight during the middle and end of the day in summer help to cool the house, and, when they lose their leaves during the winter, permit the sun's rays to warm the house. By reducing the need for air conditioning and cooling, yard trees can contribute to the conservation of fossil fuel resources and to the reduction of air pollution resulting from burning fossil fuels.

In dry climates, such as that of southern California, shade from trees can help reduce the loss of moisture through evaporation. During the rainy season, when the land can be subjected to torrential rains, trees reduce soil erosion, and aid in flood control. Trees also provide shelter from winds, which can increase evaporation, cause erosion, and produce greater cooling (the "wind-chill factor"). Many farms in prairie regions use stands of tall trees as wind breaks. Finally, trees provide an environment in which other species of plants and animals can thrive.

☞In this activity, you will make a critical assessment of the ways vegetation is used in a residential yard and suggest how it could be used to better advantage to reduce the need for heating and air conditioning.

✔MATERIALS: graph paper (5 squares/inch)
                compass (or map of the area showing north)
                yardstick

☉TIME: 1 hour preparing map, 1 hour for followup

❀ACTIVITY:

1.   Select a private yard, preferably one with a single dwelling. Measure the length of one pace (the longest step you can take) and then pace off the length and width of the yard. Multiply the number of paces by the length of one pace to obtain the approximate size of the yard. In the

same manner, measure the location of the house, trees, shrubs, lawn, and flower beds. Use your measurements to make a scale drawing of the yard.

2.   Compare the locations and types of plantings with those shown in the diagrams on this and the next page. List, in one column, those plantings that are well placed and, in another column, those that are poorly placed.

✎**FOLLOWUP:**

Prepare a proposal for modifying the plantings in the yard to improve the environment of the house.

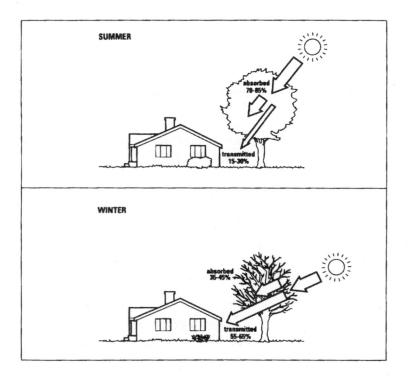

Shade and Energy Use

Source: G. M. Heisler, 1986, Energy Savings With Trees, *Journal of Arboriculture, 12*(5), in H. Akbari, et al.,1992, *Cooling Our Communities: A Guidebook on Tree Planting and Light-Colored Surfacing*, U.S. EPA Office of Policy Analysis, U.S. Supt. of Documents, ISBN 0-16-036034-X, Washington, DC.

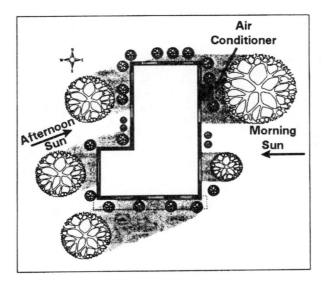

Trees Can Provide Cooling in Summer and Warming in Winter

Source: G. M.. Heisler, 1986, & J. H. Parker, 1982, in H. Akbari, H. et al., 1992, *Cooling Our Communities: A Guidebook on Tree Planting and Light-Colored Surfacing*, U.S. EPA Office of Policy Analysis, U.S. Supt. of Documents, ISBN 0-16-036034-X, Washington, DC.

Activity #41: Green Streets

Trees in urban areas have been receiving increased attention recently because they have many benefits—from taking in carbon dioxide and giving off oxygen to improving aesthetics.

The 300 million urban trees in the United States occupy 3% of the land area, or 28 million hectares (Mha). They include naturally surviving remnants of native forests, as well as planted trees, both native and exotic. Urban trees account for about 30% of the area of the average American city. The most popular planted trees are maple, oak, pine, sycamore, sweetgum, elm, ash, linden and honey locust. Generally, urbanization decreases the diversity, as well as the numbers, of trees, but in areas that were naturally sparsely wooded, the trend is the opposite. For example, Berkeley, California, which would have had few trees in pre-settlement times, has 175 species in the city and 300 on its University of California campus.

Urban trees are subjected to a variety of stressful conditions, including increased predation by insects, more disease, damage to bark, poor soil, inadequate water and daylight, high temperatures, strong winds, too much light at night, air pollution, salts from de-icing of streets and roadways, dog urine, and foot traffic.

☞In this activity, you will survey urban trees on a public street in order to understand better the stresses trees experience and to consider ways to improve the conditions of trees in an urban setting.

✔**MATERIALS:**   one pair of dice
tape measure

☺**TIME:**   1 hour for activity, 1 hour for followup

❀**ACTIVITY:**

1.   Select a city block in which trees are planted. Count the numbers and the kinds of trees on both sides of the street. If you do not know their names, collect and press leaf samples, make drawings of leaves, or take photographs.

2.   If there are more than 10 trees in the block, you should select 10 trees at random. Start at one end of the block and throw a pair of dice. If, for example, the total is 7, walk to the 7th tree. Throw the dice again to select a second tree, and so on. If there are fewer than 10 trees, include two blocks in your sample. If there are exactly 10 trees in one block, include all of them.

3.   Observe each of the 10 trees carefully for signs of stress: bark injury, resin ooze, insects, evidence of insect damage (holes in leaves, ragged or curled leaf edges, warty growths, brown spots), sparse leaves (other than that due to seasonal leaf drop), broken pavement, twisted or stunted growth, interference from buildings, wires, or other obstacles. Trees should have a minimum soil pit of 30 square feet, with a minimum width of 4 feet.

✎FOLLOWUP:

Complete a data sheet similar to the sample below. Use your conclusions to prepare a proposal for increasing the numbers and types of trees and for improving the growing conditions for the trees in the block.

SAMPLE DATA SHEET

Tree #	Bark	Resin	Leaves	Growth	Pave-ment	Relation-ship to Buildings, Wires, etc.	Area of Soil Pit (L x W) sq. ft.
1							
2							
3							
4							
5							
6							
7							
8							
9							
10							

Activity #42: Plastics by the Numbers

Plastics are now so common that it is hard to realize that they had limited uses until World War II spurred their development. Disposal of plastics was not a problem until their number and uses began multiplying dramatically in the 1950s. In 1960, plastic wastes in the United States equaled 0.4 million tons a year, but had reached 16.2 million tons by 1990. Each American disposes of one-half pound (approximately 0.2 kg) on average each day, about half of which is in the form of packaging material. Because they are not recyclable in nature, plastics do not decompose in landfills and only 2% of plastics are currently recycled. Although plastic accounts for less than 10% by weight of wastes deposited in landfills, they represent 21% by volume, due to their low density. The amount of plastic wastes is expected to increase to 20 million tons by 1995.

Petroleum and natural gas are used in the manufacture of plastics, causing concern about their impact on reserves of these fossil fuels. Manufacturing of plastics produces pollutants, including some toxic substances. Plastics are nonrenewable and degrade slowly, if at all, and plastic debris in oceans has caused the death of many thousands of marine organisms, and has also spoiled the natural beauty of beaches and open water. But plastics offer certain advantages as well: they are excellent packaging material and enable manufacturers to reduce packaging waste by about one-third. Because they are light, transportation of plastics and products packaged in plastic requires less fuel and reduces air pollution. Efforts to reduce the impact of plastics on the environment are focused on three approaches: reduce the amounts used, reuse as much as possible, and recycle the rest.

☞In this activity, you will use physical properties to characterize the common types of plastics.

✔MATERIALS: plastic containers, with number codes
            water
            sugar
            rubbing alcohol (isopropyl)
            scissors, marking pen
            2 glasses or clear plastic cups
            safety goggles

⊘**TIME:**   2 hours for activity, 1 hour for followup

❀**ACTIVITY:**

1.   Before you begin this activity, you will need to collect plastic containers with numbers 1 through 6 stamped on the bottoms. Collect containers of as many different numbers as you can. The information in the table "Common Plastics" (see page 187) may help you in your search.

2.   Cut a piece, approximately 1 inch square, from each type of plastic. Mark each piece with the number stamped on the bottom of the container from which it came. Cut the corners of the square to produce a chip that looks like this:

3.   Describe the appearance of each chip, using the characteristic pairs listed below:

          clear or opaque?
          waxy or not waxy?
          shiny or dull?
          glassy or leathery?

4.   Test the flexibility of each chip by trying to bend each with your fingers. Does it bend? bend easily? return to original position after being bent? crack when bent? partially curl? curl and spring back when released?

5.   Dissolve 5 teaspoons (25 grams) of sugar in 2.5 ounces (75 ml) of water in one of the glasses. Place a chip in the sugar solution and record whether it sinks or floats. Test each of the other chips, one at a time.

6.   Rinse the chips with water and dry thoroughly. Place 3 teaspoons (1 tablespoon or 15 ml) of isopropyl alcohol in the second glass and add enough water to make a total of 3 ounces (100 ml) of liquid. One at a time, test whether the plastic pieces float in the alcohol-water mixture.

✎FOLLOWUP:

1.   Present your results in a table, with the types of plastics in the columns and the characteristics in the rows.

2.   Explain how the characteristics of the different types could be used to separate them in a recycling program if there were no number codes.

Common Plastics			
Code No.	Type	Characteristics	Common Uses/**Products**
1	polyethylene terephthalate	tough, shatter resistant, not permeable to gas	carpeting, fiberfill, bottles (not food), containers/**soft drink and juice bottles**
2	high density polyethylene	tough, flexible, translucent	bottles for motor oil, detergent; pipes and pails/**milk jugs**
3	polyvinyl chloride	strong, clear, brittle unless treated	drainage pipes, fencing, siding for houses/**glass cleaner, shampoo, and salad dressing bottles**
4	linear low polyethylene	moisture proof, inert	trash bags, pallets/**plastic trash bags**
5	polypro-pylene	stiff, resistant to heat and chemicals	auto parts, batteries, carpeting/**yogurt and margerine containers, rigid bottle caps**
6	polystyrene	brittle, clear, rigid, good insulator	insulation board, office equipment, reusable trays/**carry-out packaging for fast foods, plastic utensils**

Source: Reprinted courtesy of Lyons and Burford, Publishers from *The Plastic Waste Primer* by the League of Women Voters Education Fund, 1993.

Activity #43: Trash Collection

The amount of solid waste generated in the United States is increasing faster than its population. Since 1960, the number of tons of solid waste has risen from 88 million to 200 million. Managing the disposal of all of this waste costs about $30 billion annually. Landfills are filling to capacity and citizens are increasingly resistant to having new ones built in their neighborhoods. But the amount of waste is only part of the problem. Disposing of materials that will not decompose drains valuable resources and wastes energy. Toxic substances leaching from landfills and contaminating soil, water, and air are serious concerns as well. While scientists and engineers continue to explore new and safer technologies for recycling and disposing of wastes, there is much that all citizens can do: reduce the amount of waste brought into their households, reuse materials or allow others to reuse them, and cooperate with community recycling programs, whether required or voluntary.

☞In this activity, you will determine how much trash you throw out in a week and the relative amounts of different types of material in the trash.

✔MATERIALS: 14 large plastic trash bags
bathroom scale or tape measure

☉TIME: 1 1/2 hours for activity; 1 hour for followup

✤ACTIVITY:

1.   Collect the trash you produce each day for a week, using one bag for each day's collection. **Do not include food wastes or any material that might be a health hazard. If you dispose of food containers, rinse them thoroughly before bagging them.**

2.   Determine the weight of each bag by first weighing yourself on the bathroom scale and then weighing yourself holding the bag. Subtract the first weight from the second to find the weight of the trash. (If you do not have access to a scale, you can use the tape measure to find the volume. To do this, form the bag into an approximation of a rectangular solid and find the length, width, and height of the bag. Multiply these to find the volume in cubic inches or cubic centimeters.)

3.   At the end of the week, add up the weights (or volumes) and divide by 7 to find the average amount of trash per day.

4.   On a daily basis or at the end of the week, sort the trash into the following categories: paper and paperboard, yard wastes, plastics, glass, metals, wood, and other, using the remaining seven bags. Tabulate the total weight (or volume) and calculate the percentage of the total trash for each category.

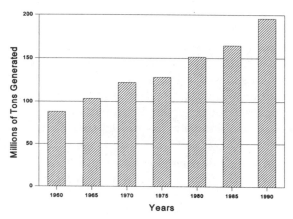

Solid Waste Generated in the U.S.

Source: Environmental Protection Agency, 1992. *Characterization of Municipal Solid Waste in the United States: 1992 Update,* United States Environmental Protection Agency.

✎FOLLOWUP:

1.   Compare your daily average to the average of 4.5 lbs. (2 kg) for all Americans. Multiply by 365 to find the amount of trash you generate in one year.

2.   Compare the composition of your trash to the national averages (excluding food wastes): paper and paperboard—40%; yard wastes—19%; metals—8.3%; plastics—9%; glass—7%; wood—7%; other—9.7%.

3.   Write a brief description of what you can do to reduce your contribution to the solid waste disposal problem. The brief list in the box on the next page may help you, but do not be limited by it.

Ways to Reduce Waste

Purchasing Power

1. If you don't need it, don't buy it.
2. Purchase products with a minimum amount of packaging, refillable containers, or ones made from recycled or recyclable materials.
3. Purchase items in the largest size or most concentrated form available.
4. Provide your own shopping bag or refuse bags whenever possible.
5. Purchase products made from recycled materials.
6. Reduce the amount of junk mail you receive. (Contact: Mail Preference Service, Direct Marketing Association, P. O. Box 3861, New York, NY 10163-3861.)

Reuse Refuse

1. Use plastic boxes, clean glass jars and coffee cans for food storage, rather than paper, plastic, or aluminum wrap.
2. Find other uses for items you would normally discard, such as using lawn clippings as garden mulch, or newspaper to clean windows.
3. Use refillable pens, rechargeable batteries, and other durable products.
4. Donate used furniture, books, magazines, toys, and clothes so that others may have use of them.

Recycle the Rest

1. Recycle aluminum cans, paper, glass, and plastics.
2. Compost kitchen waste (other than meat, cheese, grease) and yard waste.
3. Cooperate with community efforts to recycle. Follow instructions about recycling carefully—there is usually a good reason for them.

Activity #44: Bottle Composting

Soil is composed of inorganic material, from the weathering of rocks, and organic material, from the decay of plants or animals. The organic portion of soil, humus, is a source of nutrients for new plants, as well as a soil conditioner. A soil rich in humus is darker than soil with less humus. Dead organisms are fed on by a variety of plants, fungi, animals, and, eventually, by microorganisms, each participating in the progressive breakdown of the material. The final step in the production of compost is carried out by decay bacteria. Materials that can be broken down in this way are called biodegradable, whereas others, such as plastics and metals, cannot be recycled in nature. Biodegradable materials differ in the length of time they take to decompose. Newspapers, for example, although biodegradable, can last for decades in landfills, partly because they have a high cellulose content and partly because there is not sufficient air in the landfills to promote decay.

Gardeners and farmers often make use of the natural fertilizing properties of humus to supplement or replace chemical fertilizers. The humus is produced by mixing waste materials in large containers with soil, worms, and, often, some chemical fertilizer, a process known as composting. Meats, fats, and oils are not used because they attract animals. When the compost has the right mixture and sufficient air and water, it begins to decompose, as indicated by a rise in temperature of the compost to about 150°F. Many communities use composting on a large scale to recycle yard waste and sewage sludge and reduce the volume of material going into landfills. The product is used to improve poor soils, to cover landfills, and as a top dressing for lawns and fields of crops.

☞In this activity you will construct a compost column and observe what happens during the composting process.

✔MATERIALS: organic materials, such as vegetable peelings, fruit peelings and cores, eggshells, wood shavings, grass clippings, dried plants, leaves (no meat, fats, or oils should be used)
3 empty 2-liter plastic soda bottles
scissors
single-edge razor blade or sharp knife

water
marking pen
large sharp needle or large nail
candle
garden soil or potting soil (not sterile)
lawn fertilizer or liquid plant "food"
trowel or other implement, if using garden soil
earthworms, if available
stick or other implement about as long as the
      compost column

☙TIME: 2 1/2 hours for activity; 1/2 hour for followup

❀ACTIVITY:

1.   Remove the labels from the bottles using hot water. With the marking pen, mark a ring around one bottle halfway between the top and bottom. Cut the bottle (1) in half (see diagram, next page).

2.   Mark and cut a second bottle just before the sides taper towards the base. You want the top piece to end with straight sides. Separate the rounded bottom of the bottle (2) from the plastic base.

3.   Mark and cut the third bottle (3) in two places: (a) toward the top, just after the sides taper inward, and (b) towards the base, just after the sides taper inward. You want the middle piece to taper somewhat at both cut ends.

4.   Heat the needle in the flame of the candle and use it to poke a number of holes in the sides of the pieces of the bottles. Cover the spout of the second bottle with a piece of netting or nylon stocking and hold it in place with a rubber band. Chop the waste organic material into small pieces.

5.   Assemble your column as follows, from top to bottom:

   a.   the bottom half of bottle #1
   b.   the top of bottle #2, with the spout end down
   c.   the middle portion of bottle #3 (the tapered ends should fit neatly into the straight sides of bottle #2)
   d.   the top (spout end) of bottle #3, without its cap.

6.   Fill the compost column with alternating layers of dirt (1 inch) and organic waste (2 inches), finishing with dirt. Sprinkle each layer with a small amount of water and fertilizer. (If using liquid fertilizer, mix with water according to directions.) The material should feel damp, but not soggy. No water should drip into the base of the column. If you add earthworms, put them in the top layer of soil.

7.   Place the column in a dark or shady place where the temperature will not vary greatly. Once a week, use a stick or other implement to stir the contents of the column. If any water drains into the base of the column, recycle it by pouring it into the top of the column.

✎FOLLOWUP:

1.   Make a record of the materials you put into the compost column and its appearance on the day you set it up. Make additional observations every third day for several weeks, or until you can see evidence that composting is occurring. Describe the final product.

2.   Why does the temperature of composting material rise? Why is air necessary for composting to occur?

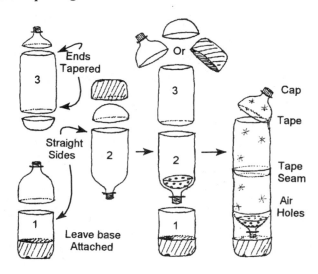

Source: © Bottle Biology Project (funded by the National Science Foundation), 1991, *Bottle Biology*, Madison, WI: Department of Plant Pathology, University of Wisconsin. Reprinted with permission.

Activity #45: Risk Assessment

Environmental decisions involve an analysis of risks that affect humans and other living organisms. In a democracy, citizens participate in the decision-making process. Therefore, the ways in which people perceive risks, and the differences in perception among nonexperts, experts, and lawmakers, are critical to effective communication, decision-making, and management related to the environment. Risk assessment, however, is always based on a degree of uncertainty, and how to manage risk is based on values and opinions, as well as scientific observations. Interpretation of data and recommendations on courses of action are often controversial. When information is incomplete, the usual procedure in science would be to seek more information. In risk management, however, decisions often need to be made before more information can be obtained. To wait for complete information is, itself, a risky procedure.

☞In this activity, you will compare your perceptions of risks with those of experts and analyze the differences.

⊘TIME: 1 hour for activity, 1 hour for followup

❀ACTIVITY:

1.   Study the list of potential risks on the next page and rank them from 1 to 12, with 1 being the greatest risk, and 12 being the least risk.

2.   Rate each risk on two scales: the observable-not observable scale and the controllable-uncontrollable scale. For each scale, use the numbers from 1 to 5 to rate each risk, with 5 being the most observable or controllable. Use the explanations of these terms that follow to decide how to rate each risk.

> **Observable risks** are those that are known to you, have an immediate effect that you can observe, are old (i.e., familiar) risks, and are known to science. **Not observable risks** are those that you are unaware of, have a delayed effect that you cannot observe immediately, are new risks, and are relatively unknown to science.

Controllable risks are those that you feel you have some control over, do not give you a feeling of "dread," are not global catastrophes, do not have fatal consequences, are fairly distributed among groups of people, pose a low risk to future generations, have an element of risk that is easily reduced or is currently decreasing, and are incurred voluntarily. **Uncontrollable risks** are those that cause you to feel dread, are global catastrophes, are often fatal, are not fairly distributed, pose a high risk to future generations, have an element of risk that is not easily reduced or is increasing, and are incurred involuntarily.

Potential Risk Factors

observable- not observable	risk factor	controllable- uncontrollable
1 2 3 4 5	__motorcycles	1 2 3 4 5
1 2 3 4 5	__pesticides	1 2 3 4 5
1 2 3 4 5	__commercial aviation	1 2 3 4 5
1 2 3 4 5	__handguns	1 2 3 4 5
1 2 3 4 5	__oral contraceptives	1 2 3 4 5
1 2 3 4 5	__X rays	1 2 3 4 5
1 2 3 4 5	__nuclear power	1 2 3 4 5
1 2 3 4 5	__alcoholic beverages	1 2 3 4 5
1 2 3 4 5	__antibiotics	1 2 3 4 5
1 2 3 4 5	__motor vehicles	1 2 3 4 5
1 2 3 4 5	__smoking	1 2 3 4 5
1 2 3 4 5	__electric power (nonnuclear)	1 2 3 4 5

Source: Adapted from "Ranking Risks: Reality and Perception," *The New York Times*, Feb. 1, 1994, p. C10. Copyright © 1994 by The New York Times Company. Reprinted by permission.

✎FOLLOWUP:

1.   Compare your ranking of the 12 risks with those of experts (see Appendix). What differences are there? Can you explain the differences by studying the ways you rated the risks on the two scales?

2.   Compare the ranking of the 12 risks by the nonexperts and the experts. To which ranking is yours most similar? What can you conclude in general about the ways in which people assess risks?

Steps in Risk Assessment		
Step	**Guiding Question**	**Example**
hazard identification	Is there a cause and effect relationship between a factor and harm to people or the environment?	Does smoking cause lung cancer?
dose-response relationship	What is the quantitative relationship betweeen the factor and harm?	Smoking 10 cigarettes a day increases the chance of getting lung cancer 25 times, on the average.
exposure analysis	Given the dose-response for known populations, what is the predicted response for other populations?	Response is linear, so that 20 cigarettes a day will increase the chances of lung cancer 50 times.
risk characterization	Given the information from the first three steps, what is the overall risk?	The overall risk is that smoking increases the chances of getting lung cancer in proportion to the number of cigarettes smoked.

Activity #46: Environmental Attitudes

Citizens can help shape environmental policies and affect implementation of those policies by electing legislators and executives who are concerned about the environment and by cooperating with environmental regulations and programs. For citizens to fulfill their environmental roles effectively and responsibly, they need to have positive attitudes toward the environment, as well as a basic understanding of how the environment works. The activities in this book are designed to help people like yourself understand the fundamental concepts in environmental science. But how do you and others like you feel about environmental issues? And, how do these attitudes compare with those of the majority of Americans? What are the trends in attitudes toward the environment? Finally, do positive attitudes translate into responsible actions?

☞In this activity, you will compare attitudes toward the environment of yourself and those around you with opinions expressed in a national poll.

☺TIME: 1 hour for activity, 1 hour for followup

✿ACTIVITY:

Answer the questions in the "Survey of Environmental Attitudes" (beginning on next page) yourself. Then, prepare a table for recording the answers of at least 10 people, selected at random, to the survey. Administer the survey to them and record their answers.

✎FOLLOWUP:

1.   Calculate the percentages for the various responses to the questions and compare your results with those from the 1991 Gallup Poll (see Appendix). Discuss the possible reasons for differences between the two.

2.   What conclusions can you draw about attitudes toward the environment from your survey and the Gallup Poll? Do people's attitudes about problems that affect them indirectly differ from those that affect them directly? Do people's attitudes about problems in other areas of the world differ from those in this country? Do positive attitudes toward the environment appear to translate into responsible actions?

Survey of Environmental Attitudes[1]

1.   Do you consider yourself to be an environmentalist or not?

_____ yes   _____ no   _____ no opinion

2.   (If yes to #1): Would you say you are a strong environmentalist or not?

_____ yes   _____ no   _____ no opinion

3.   How much progress have we made in dealing with environmental problems in this country in the past 20 years? Would you say we have made a great deal of progress, only some progress, or hardly any progress at all?

_____ great deal   _____ only some

_____ hardly any   _____ no opinion

4.   How much optimism do you have that we will have our environmental problems well under control in 20 years time?

_____ a great deal of optimism   _____ only some optimism

_____ hardly any optimism   _____ no opinion

5.   Do you think the American public is too worried about the environment, not worried enough, or expresses about the right amount of concern about the environment?

_____ too worried   _____ not enough worried

_____ right amount of concern   _____ no opinion

6.   Do you think the government is too worried about the environment, not worried enough, or expresses about the right amount of concern about the environment?

---

[1] Questions have been taken from G. Hueber, 1991 (Apr.), Americans Report High Levels of Environmental Concern, Activity, *The Gallup Poll Monthly*, pp. 6-12. Reprinted by permission of the Gallup Poll Organization.

_____ too worried   _____not enough worried

_____ right amount of concern   _____ no opinion

7. Do you think business and industry are too worried about the environment, not worried enough, or express about the right amount of concern about the environment?

_____ too worried   _____not enough worried

_____ right amount of concern   _____ no opinion

8. Here are two statements which people sometimes make when discussing the environment and economic growth. Which of these statements comes closer to your own point of view?

_____ Protection of the environment should be given priority, even at the risk of curbing economic growth.

_____ Economic growth should be given priority, even if the environment suffers to some extent.

_____ No opinion.

9. For each of the environmental problems that follow, please state whether you personally worry about them a great deal, a fair amount, only a little, or not at all?

Problem	a great deal	a fair amount	only a little	not at all	no opinion
drinking water	_____	_____	_____	_____	_____
pollution of rivers, lakes, reservoirs	_____	_____	_____	_____	_____
toxic waste	_____	_____	_____	_____	_____
air pollution	_____	_____	_____	_____	_____

ocean and beach pollution	____	____	____	____	____
loss of natural habitat for wildlife	____	____	____	____	____
damage to the ozone layer	____	____	____	____	____
radioactive contamina-tion from nuclear facilities	____	____	____	____	____
loss of tropical rain forests	____	____	____	____	____
global warming	____	____	____	____	____
acid rain	____	____	____	____	____

10.   Which of the following things, if any, have you or other household members done in recent years to try to improve the quality of the environment?

Action	Yes	No/No Opinion
voluntarily recycled newspapers, glass, aluminum, motor oil, or other items	____	____
avoided buying or using aerosol sprays	____	____
contributed money to an environmental, conservation, or wildlife preservation group	____	____
specifically avoided buying a product because it was not recyclable	____	____

used cloth rather than disposable diapers   _____   _____

boycotted a company's products because of its   _____   _____
record on the environment

did volunteer work for an environmental,   _____   _____
conservation, or wildlife preservation group

cut your household's use of water   _____   _____

cut your household's use of energy by   _____   _____
improving insulation or changing your
heating or air conditioning system

replaced a "gas guzzling" automobile with   _____   _____
one that is more fuel efficient

cut down on the use of a car by carpooling or   _____   _____
taking public transportation

APPENDIX

Making an Indicator Solution

✔**MATERIALS:**   1 head of red cabbage, approximately 1 pound
water
large cooking pot
measuring cup
1 quart jar, with screw type lid

INSTRUCTIONS:

1.   Cut the cabbage into large wedges and put it in the cooking pot. Add water to cover. Bring to a boil and cook for 20 minutes.

2.   Drain liquid into the jar. When the liquid is cool, cover the jar and put it into the refrigerator. The liquid should be a dark purple. Discard the cabbage.

3.   This indicator will keep, if refrigerated, for about 1 week. If it changes color, discard and make a fresh batch.

Potential Risks Ranked by Experts and Nonexperts

Nonexperts	Potential Risk	Experts
2	motor vehicles	1
4	smoking	2
6	alcoholic beverages	3
3	handguns	4
5	motorcycles	5
11	X rays	6
7	pesticides	7
9	electric power (nonnuclear)	8
10	oral contraceptives	9
8	commercial aviation	10
1	nuclear power	11
12	antibiotics	12

Source: Adapted from "Ranking Risks: Reality and Perception," *The New York Times*, Feb. 1, 1994, p. C10. Copyright © 1994 by The New York Times Company. Reprinted by permission.

Survey of Environmental Attitudes[1]

1.　Do you consider yourself to be an environmentalist or not?

　　<u>78%</u>　yes　　　　<u>19%</u>　no　　　　<u>3%</u>　no opinion

2.　(If yes to #1): Would you say you are a strong environmentalist or not?

　　<u>37%</u>　yes　　　　<u>39%</u>　no　　　　<u>2%</u>　no opinion

3.　How much progress have we made in dealing with environmental problems in this country in the past 20 years? Would you say we have made a great deal of progress, only some progress, or hardly any progress at all?

　　<u>18%</u>　great deal　　　　　　　<u>62%</u>　only some

　　<u>19%</u>　hardly any　　　　　　　<u>1%</u>　no opinion

4.　How much optimism do you have that we will have our environmental problems well under control in 20 years time?

　　<u>19%</u>　a great deal of optimism　　　　<u>60%</u>　only some optimism

　　<u>18%</u>　hardly any optimism　　　　　<u>3%</u>　no opinion

5.　Do you think the American public is too worried about the environment, not worried enough, or expresses about the right amount of concern about the environment?

　　<u>6%</u>　too worried　　　　　　<u>70%</u>　not enough worried

　　<u>22%</u>　right amount of concern　　<u>2%</u>　no opinion

---

[1] Questions and data have been taken from G. Hueber, 1991 (Apr.), Americans Report High Levels of Environmental Concern, Activity, The Gallup Poll Monthly, pp. 6-12. Reprinted by permission of the Gallup Poll Organization.

6.   Do you think the government is too worried about the environment, not worried enough, or expresses about the right amount of concern about the environment?

   <u>3%</u>   too worried   <u>73%</u> not enough worried

   <u>20%</u> right amount of concern   <u>4%</u> no opinion

7.   Do you think business and industry are too worried about the environment, not worried enough, or express about the right amount of concern about the environment?

   <u>1%</u>   too worried   <u>83%</u> not enough worried

   <u>13%</u> right amount of concern   <u>3%</u>   no opinion

8.   Here are two statements which people sometimes make when discussing the environment and economic growth. Which of these statements comes closer to your own point of view?

   <u>71%</u>   Protection of the environment should be given priority, even at the risk of curbing economic growth.

   <u>20%</u>   Economic growth should be given priority, even if the environment suffers to some extent.

   <u>9%</u> No opinion.

9.   For each of the environmental problems that follow, please state whether you personally worry about them a great deal, a fair amount, only a little, or not at all?

Problem	a great deal	a fair amount	only a little	not at all	no opinion
drinking water	67%	19%	10%	3%	1%
pollution of rivers, lakes, reservoirs	67%	21%	8%	3%	1%
toxic waste	62%	21%	11%	5%	1%

air pollution	59%	27%	10%	4%	<1%
ocean and beach pollution	53%	26%	14%	6%	1%
loss of natural habitat for wildlife	53%	27%	15%	5%	<1%
damage to the ozone layer	49%	24%	15%	8%	4%
radioactive contamination from nuclear facilities	44%	25%	20%	10%	1%
loss of tropical rain forests	42%	25%	21%	10%	2%
global warming	35%	27%	21%	12%	5%
acid rain	34%	30%	20%	13%	3%

10. Which of the following things, if any, have you or other household members done in recent years to try to improve the quality of the environment?

Action	Yes	No/No Opinion
voluntarily recycled newspapers, glass, aluminum, motor oil, or other items	86%	14%
avoided buying or using aerosol sprays	68%	32%

contributed money to an environmental, conservation, or wildlife preservation group	51%	49%
specifically avoided buying a product because it was not recyclable	49%	51%
used cloth rather than disposable diapers	25%	75%
boycotted a company's products because of its record on the environment	28%	72%
did volunteer work for an environmental, conservation, or wildlife preservation group	18%	82%
cut your household's use of water	68%	32%
cut your household's use of energy by improving insulation or changing your heating or air conditioning system	73%	27%
replaced a "gas guzzling" automobile with one that is more fuel efficient	67%	33%
cut down on the use of a car by carpooling or taking public transportation	46%	54%

SOURCES OF INFORMATION AND MATERIALS

Sources of Information and Materials

1.   For further information on bird feeding and other activities involving birds, contact Cornell Lab of Ornithology, 150 Sapsucker Woods Road, Ithaca, NY 14850 (Tel: 607-254-2440). The lab is an international center for the study, appreciation, and conservation of birds, which sponsors public outreach programs (such as Project Feeder Watch, Project Pigeon Watch, Project Tanager, and the Seed Preference Test), as well as research on birds and bird communication.

2.   Owl pellets can be obtained from: Genesis, Inc., P. O. Box 2242, Mount Vernon, WA 98273; Pellets, Inc., P. O. Box 5484, Bellingham, WA 98227; and from most general biological supply companies.

3.   A good source of information on population-related issues is the Population Reference Bureau, Inc.,1875 Connecticut Ave., NW, Suite 520, Washington, DC 20009-5728. The Population Reference Bureau is a private, nonprofit organization dedicated to objective analysis and reporting of population issues that affect the U.S. and the world. The Bureau publishes a monthly newsletter, *Population Today*, quarterly *Population Bulletins* on special topics, and annual *World and U.S. Population Data Sheets*.

4.   The 4-H Program is an informal education program for youth 9-19 years old. It provides a variety of experiences to help youth become skilled and responsible adults. For information on the 4-H Program, contact the National 4-H Council, 7100 Connecticut Avenue,Chevy Chase, MD 20815.

        If you are interested in learning more about the Cornell Cooperative Extension 4-H Youth Development Program, contact Cornell Cooperative Extension, Education Center, 16 East 34th Street, 8th floor, New York, NY 10016-4328.

CREDITS

Credits

Activity #3: Bird Food

Adapted from the "Seed Preference Test," Cornell Lab of Ornithology, Ithaca, New York, by permission.

Activity #7: Estimating Biodiversity

© The Ohio State University. Adapted from Activities for the Changing Earth System, Earth Systems Education Program, 059 Ramseyer Hall, 29 W. Woodruff Avenue, Columbus, OH 43210.

Activity #11: Improving a Wildlife Habitat

Adapted from National Institute for Urban Wildlife. 1991. *Developed Lands: Restoring and Managing Wildlife Habitats* (Teacher's Pac series in Environmental Education). National Institute for Urban Wildlife, P. O. Box 3015, Shepherdstown, WV 25443.

Activity #14: Food Web Interactions

Based on information in Spencer, C.N., R. McClelland, & J. A. Stanford, 1991, Shrimp Stocking, Salmon Collapse, and Eagle Displacement, *Bioscience*, *41*(1):14-21.

Activity #18 : Carbon Budget

Adapted from D. J. Wadington, & H. M. Gardner, 1991, *Education in Global Change: An Interdisciplinary Science Project, by the Committee on Teaching Science*, International Council of Scientific Unions.

Activity #24: Energy and Agriculture

Adapted by permission from National Science Teachers Association, 1980, T*he Energy Future Today—Grades 7, 8, 9 Social Studies*, Washington, DC, National Science Teachers Association. Copyright ©1980 by the National Science Teachers Association.

Activity #25: Wind Power

Adapted by permission from New York Energy Education Project,1988, *Renewable Energy: Student Activities*, Albany, NY: The Energy Project.

Activity #27: Radioactive Decay

Adapted from John Lord, *Energy from Nuclear Reactions. Teaching About Energy, Part Four.* Copyright © 1987 by Enterprise for Education, Inc. Reprinted by permission.

Activity #28: Radiation Exposure

Adapted from John Lord, *Energy from Nuclear Reactions. Teaching About Energy, Part Four.* Copyright © 1987 by Enterprise for Education. Reprinted by permission.

Activity #29: Water, Water, Everywhere?

Part A adapted with permission from *Earth: The Water Planet.* Copyright © 1989, 1992 by the National Science Teachers Association.

Activity #33: Stream Quality Indicators

Adapted from Division of Natural Areas and Preserves, Scenic Rivers Section, 1993, *A Guide to Volunteer Stream Quality Monitoring*, Ohio Department of Natural Resources, Columbus, OH, 23 pp.

Activity #34: Lichen Indicators

Adapted and modified from "Lichen Looking," from the *Outdoor Biology Instructional Strategies (OBIS) series.* Copyright © The Regents of the Unversity of California. Available from Lawrence Hall of Science, University of California at Berkeley.

Activity #35: Chemical Pollutants

Adapted from B. S. Kendler, & H. G. Koritz, 1990, Using the Allium Test to Detect Environmental Pollutants, *The American Biology Teacher*, 52(6): 372-375.

Activity #36: Greenhouse Globe

Adapted and modified from *Global Warming and the Greenhouse Effect*, one of more than 40 teacher's guides in the *Great Explorations in Math and Science (GEMS)* series, available from the Lawrence Hall of Science, University of California at Berkeley.

Activity #42: Plastics by the Numbers

Adapted by permission, from "How Can Plastics Be Separated?" in *Solid Waste: Is There A Solution?*, New York State Science, Technology, and Society Education Program, 1992.

Activity #44: Bottle Composting

Adapted by permission, from "Compost Columns," *Bottle Biology*, © 1991, Department of Plant Pathology, University of Wisconsin, 1630 Linden Drive, Madison, WI 53706.

REFERENCES

References

Activity #1: People and Nature
Botkin, D. B. 1990. *Discordant Harmonies*. New York: Oxford University.
Oelschlaeger, M. 1991. *The Idea of Wilderness: From Prehistory to the Age of Ecology*. New Haven: Yale University.

Activity #2: Earth Systems
Barlow, C., & T. Volk. 1992 (Oct.). Gaia and Evolutionary Biology. *Bioscience*, *42*(9): 686-692.
Lovelock, J. E. 1982. *Gaia: A New Look at Life on Earth*. New York: Oxford University.
Margulis, L., & J. E. Lovelock. 1989. Gaia and Geognosy. In M. B. Rambler, L. Margulis, & R. Fester (eds.), *Global Ecology: Towards a Science of the Biosphere*, San Diego, CA: Academic.
Schneider, S. H. 1990 (May). Debating Gaia. *Environment, 32*(4): 4-9, 29-32.

Activity #3: Bird Food
Grubb, T. C. 1986. *Beyond Birding: Field Projects for Inquisitive Birders*. Pacific Grove, CA: Boxwood.

Activity #4: Plant Seedlings
McCain, G., & E. M. Segal. 1973. *The Game of Science*. Monterey, CA: Brooks/Cole.

Activity #5: The Power of Doubling
Population Reference Bureau. 1993. *World Population Data Sheet*. Washington, DC: Population Reference Bureau, Inc.
Kent, M. M., & K. A. Crews. 1990. *World Population: Fundamentals of Growth*. Washington, DC: Population Reference Bureau, Inc.

Activity #6: *Mark and Capture*
Schmidt, K. 1994 (Jan. 4). Scientists Count a Rising Tide of Whales in the Seas. *Science, 263*: 25-26.
Southwood, T. R. E. 1966. *Ecological Methods: With Particular Reference to the Study of Insect Populations*. London: Methuen.
Whitney, C. R. 1992 (June 30). Norway Is Planing to Resume Whaling Despite World Ban. *New York Times*, A1, A2.
Wratten, S. D. & G-L. A. Fry. 1980. *Field and Laboratory Exercises in Ecology*. London: Edward Arnold.

Activity #7: Estimating Biodiversity
Ehrlich, P. R., & E. O. Wilson. 1991(Aug.). Biodiversity Studies: Science and Policy. *Science, 253*: 758-761.
Erwin, T. L. 1991 (Sept.). How Many Species Are There?: Revisited. *Conservation Biology, 5*(3): 330-333.

Gaston, K. J. 1991 (Sept.) The Magnitude of Global Species Richness. *Conservation Biology, 5*(3): 283-296.

Wilson, E. J. (Ed.). 1988. *Biodiversity.* Washington, DC: National Academy.

Wolf, E. C. 1991. Survival of the Rarest. *World Watch, 4*(2): 12-20.

Activity #8: Making an Age Pyramid
McFalls, J.A., Jr. 1991 (Oct.). Population: A Lively Introduction. *Population Bulletin, 46*(2).

Population Reference Bureau. 1972. *The World Population Dilemma.* Washington, DC: Population Reference Bureau, Inc.

Activity #10: Wildlife in Your Area
Krasny, M.E. n.d. *Wildlife in Today's Landscapes.* Ithaca, NY: Cornell Cooperative Extension.

National Institute for Urban Wildlife. n.d. *Urban Areas* (Wildlife Habitat Conservation Teacher's Pac Series). Columbia, MD: National Institute for Urban Wildlife.

Van Druff, L. 1979 (Jul.-Aug.). The Wildlife in Your Backyard. *The Conservationist, 34*(1): 9-11.

Activity #11: Improving a Wildlife Habitat
National Institute for Urban Wildlife. 1991. *Developed Lands: Restoring and Managing Wildlife Habitats* (Habitat Pac). Columbia, MD: National Institute for Urban Wildlife.

Activity #12: Community Relations
Pratt, C. 1987. Gray Squirrels As Subjects in Independent Study. *The American Biology Teacher, 49*(8): 434-437.

Activity #13: Creating a Mini-Ecosystem
Appenzeller, T. 1994 (March 11). Biosphere 2 Makes a New Bid for Scientific Credibility. *Science, 263*: 1368-1369.

McCourt, R. 1988 (Jan./Feb.). Creating Miniature Worlds. *International Wildlife, 18*: 38-40.

Nelson, M., T. L. Burgess, A. Alling, N. Alvarez-Romo, W. F. Dempster, R. L. Walford, & J. P. Allen. 1993 (April). Using a Closed Ecological System to Study Earth's Biosphere. *Bioscience, 43*(4): 225-237.

Activity #14: Food Web Interactions
Spencer, C.N., R. McClelland, & J. A. Stanford. 1991. Shrimp Stocking, Salmon Collapse, and Eagle Displacement, *Bioscience, 41*(1): 14-21.

Activity #15: Owl Pellets
Forbes, G. 1992 (Jan.). White Ghost of the Arctic. *Wildbird*, 30-33.

Genesis. 1989. *Owl Pellet Pak: Instructor's Guide.* Mount Vernon, WA: Genesis

Litchfield, M. 1992 (Jan.). Day (and Night) of the Owl. *Wildbird*, 62-63.

Marti, C. D. 1988. The Common Barn-Owl. *Audubon Wildlife Report: 1988*. New York: Academic, 535-550.

Marti, C. D. 1973. Food Consumption and Pellet Formation Rates in Four Owl Species. *Wilson Bulletin, 85*(2): 178-181.

Raptor Trust. n.d. *Owl Facts: Focusing on N. J.'s Owls*. Millington, N. J.: The Raptor Trust.

Szafranski, K. 1992 (Jan.). Owl Encounters. *Wildbird*, 34-39.

Whitaker, J. O., Jr. 1980. *The Audubon Society Field Guide to North American Mammals*. New York: Knopf.

Activity #16: Carrying Capacity
Goodrum, P. D., V. H. Reid, & C. E. Boyd. 1971(July). Acorn Yields, Characteristics, and Management Criteria of Oaks for Wildlife. *Journal of Wildlife Management, 35*(3): 520-532.

Feldhammer, G. A., T. P. Kilbane, & D. W. Sharp. 1989. Cumulative Effect of Winter on Acorn Yield and Deer Body Weight. *Journal of Wildlife Management, 53*(2): 292-295.

Harlow, R. F., J. B. Whelan, H. S. Crawford & J. E. Skeen. 1975. Deer Foods During Years of Oak Mast Abundance and Scarcity. *Journal of Wildlife Management, 39*(2): 330-336.

McShea, W. J., & G. Schwede. 1993. Variable Acorn Crops: Responses of White-Tailed Deer and Other Mast Consumers. *Journal of Mammology, 74*(4): 999-1006.

Petrides, G. A. 1958. *A Field Guide to Trees and Shrubs*. Boston: Houghton Mifflin.

Activity #17: Plant Productivity
Coleman, D. C., & P. F. Hendrix. 1988. Agroecosystems Processes. In L. R. Pomeroy, & J. J. Alberts (eds.). *Concepts of Ecosystem Ecology*. Amsterdam: Springer-Verlag, 149-170.

Activity #18: Carbon Budget
Berner, R. A., & A. C. Lasaga. 1989 (March). Modeling the Geochemical Carbon Cycle. *Scientific American, 260*(3): 74-81.

Dixon, R. K., S. Brown, R. A. Houghton, A. M. Solomon, M. C. Trexler, & J. Wisniewski. 1994 (Jan. 14). Carbon Pools and Flux of Global Forest Ecosystems. *Science, 263*: 185-190.

Moore, B., M. P. Gilden, C. J. Vorasmaty, D. L. Skole, J. M. Melillo, B. J. Peterson, E. B. Rastetter, & P. A. Stendler. 1989. Biogeochemical Cycles. In M. B. Rambler, L. Margulis, & R. Fester. *Global Ecology: Towards A Science of the Biosphere*. San Diego, CA: Academic, 113-141.

Sundquist, E. T. 1993 (Feb. 12). The Global Carbon Dioxide Budget. *Science, 259*: 934-941.

Zimmer, C. 1993 (Dec.). The Case of the Missing Carbon. *Discover*, 38-39.

Activity #19: Litter Bugs
McCourt, R. 1988 (Jan./Feb.). Creating Miniature Worlds. *International Wildlife, 18*: 38-40.

Activity #20: A Lot of Rot
Murphy, P. 1993. *The Garbage Primer*. Washington, DC: League of Women Voters Education Fund.

Activity #21: Succession
Benton, A. H., & W. E. Werner, Jr. 1965. *Manual of Field Biology and Ecology*. Minneapolis, MN: Burgess.

Brewer, R., & M. T. McCann. 1982. *Laboratory and Field Manual of Ecology*. Philadelphia: Saunders College.

Brower, J. E., J. H. Zar, & C. N. von Ende. 1990. *Field and Laboratory Methods for General Ecology* (3rd ed.). Dubuque, IA: Wm. C. Brown.

Jones, S. B., D. de Calestra, & S. Chunko. 1993 (Nov./Dec.). Whitetails are changing our woodlands. *American Forests, 99*(11 & 12): 20-25, 53-54.

Lederer, R. J. 1984. *Ecology and Field Biology*. Menlo Park, CA: Benjamin/Cummings.

Activity #22 : Sunlight Becomes You
Broad, W. J. 1993 (Dec. 28). Strange New Microbes Hint at a Vast Subterranean World. *New York Times*, C1,C14.

Activity #23: Fossil Fuels
Flavin, C. 1992. Building a Bridge to Sustainable Energy. In L. R. Brown et al. (eds.), *State of the World, 1992*. New York: Norton.

World Resources Institute. 1991. *The 1992 Information Please Environmental Almanac*. Boston: Houghton Mifflin.

Activity #24: Energy and Agriculture
Coleman, D. C., & P. F. Hendrix. 1988. Agroecosystems Processes. In L. R. Pomeroy, & J. J. Alberts (eds.). *Concepts of Ecosystem Ecology*. Amsterdam: Springer-Verlag, 149-170.

Ghersa, C. M., M. L. Roush, S. R. Radosevich, & S. M. Cordray. 1994. Coevolution of Agroecosystems and Weed Management. *Bioscience, 44*(2): 85-94.

National Science Teachers Association. 1979. *Agriculture, Energy, and Society*. Washington, DC: National Science Teachers Association.

Activity #25: Wind Power
Brower, M. 1992. *Cool Energy*. Cambridge, MA: MIT Press.

Brown, L. R., H. Kane, & E. Ayres. 1993. *Vital Signs: 1993*. NewYork: Norton.

Gipe, P. 1993. *Wind Power for Home and Business*. Post Mills, VT: Chelsea Green

Wald, M. L. 1994 (April 12). Cheap Electricity Stalls Wind Power as an Energy Source. *New York Times*, B2.

Activity #26: Solar Still
Brower, M. 1992. *Cool Energy*. Cambridge, MA: MIT Press.
Brown, L. R., H. Kane, & E. Ayres. 1993. *Vital Signs: 1993*. New York: Norton.
Callahan, S. 1986. *Adrift*. New York: Ballantine.
Grolier, Inc. 1992. *Multimedia Encyclopedia Version I*. Software Toolworks, Inc.

Activity #28: Radiation Exposure
Goleman, D. 1994 (Feb. 1). Assessing Risk: Why Fear May Outweigh Harm. *New York Times*, C1,C10.
League of Women Voters Education Fund. 1993. *The Nuclear Waste Primer*. New York: Lyons and Burford.
Lean, G., D. Hinrichsen, & A. Markham. 1990. *Atlas of the Environment*. New York: Prentice Hall.

Activity #29: Water, Water, Everywhere?
Falkenmark, M. & C. Widstrand. 1992 (Nov.) Population and Water Resources: A Delicate Balance. *Population Bulletin*, *47*(3): 36pp.
World Resources Institute. 1992. *Information Please Environmental Almanac*. Boston: Houghton Mifflin.

Activity #30: Living Resources
Breisch, A. 1993-94 (Winter). In Search of Coldblooded Wildlife. *Wild in New York*, (Department of Environmental Conservation, Division of Fish and Wildlife Newsletter).
Ehrlich, P. R., & A. H. Ehrlich. 1990. *The Population Explosion*. New York: Simon & Schuster.
Pimm, Stuart. 1994 (May/June). What the Woods Won't Whisper. *The Sciences*, *34*(3):16-19.
Terborgh, J. 1989. *Where Have All the Birds Gone?* Princeton, NJ: Princeton University.
Wilson, E. O. 1992. *The Diversity of Life*. New York: Norton.

Activity #31: Food for Thought
_____. 1993 (April/June). Resource Rich and Poor. *EPA Journal, 19* (2): 8-9.
Bongaarts, J. 1994 (March). Can the Growing Human Population Feed Itself? *Scientific American, 270* (3): 36-42.
Durning, A. 1991(May/June). Fat of the Land. *World Watch*, 1-17.
Gershoff, S. 1990. *The Tufts University Guide to Total Nutrition*. New York: Harper & Row.
National Research Council. 1989. *Recommended Dietary Allowances* (10th edition). Washington, DC: National Academy.
Postel, S. 1994. Carrying Capacity: Earth's Bottom Line. In L. R. Brown et al. (eds.), *State of the World, 1994*. New York: Norton.

Activity #32: Land Use

Bowman, M. L., & J. F. Disinger. 1977 (June). *Land Use Management Activities for the Classroom*. Columbus, OH: ERIC Center.

Daugherty, A. B. 1991. *Major Uses of Land in the United States: 1987* (Agricultural Economic Report #643). Washington, DC: U.S. Department of Agriculture.

Houghton, R. A. 1994 (May). The Worldwide Extent of Land-use Change. *Bioscience*: *44*(5): 305-312.

World Resources Institute. 1992. *World Resources: 1992-93*. New York: Oxford University.

Yasso, W. E., & P. W. Morgan,. 1991/1992 (Dec./Jan.). Land-Use Planning: A Project for Earth- and Environmental-Studies Classes. *Journal of College Science Teaching,* 159-163.

Activity #33: Stream Quality Indicators

Caduto, M. J. 1985. *Pond and Brook: A Guide to Nature in Freshwater Environments*. Hanover and London: University Press of New England.

Mitchell, M. K., & W. B. Stapp. 1993. *Field Manual for Water Quality Monitoring*. Dexter, MI: Thomson-Shore.

Ohio Department of Natural Resources. 1993. *A Guide to Volunteer Stream Quality Monitoring*. Columbus, OH: Ohio Department of Natural Resources.

Activity #34: Lichen Indicators

Anderson, F. K., & M. Treshow. 1984. Responses of Lichens to Atmospheric Pollution. In M. Treshow (ed.) *Air Pollution and Plant Life*, New York: Wiley.

Brodo, I. M. 1971. Lichens and Air Pollution. *The Conservationist, 26*(1): 22-26.

Ferry, B. W., M. S. Baddeley, & D. C. Hawksworth. 1973. *Air Pollution and Lichens*. Toronto: University of Toronto.

Hale, M. E. 1969. *How To Know the Lichens*. Dubuque, IA: Wm. C. Brown.

Hawksworth, D. L., & F. Rose. 1976. *Lichens As Pollution Monitors*. London: Edward Arnold.

Activity #35: Chemical Pollutants

Global Tomorrow Coalition. 1990. *The Global Ecology Handbook*. Boston: Beacon.

Misch, A. 1994. Assessing Environmental Health Risks. In L. R. Brown et al. (eds.), *State of the World: 1994*, New York: Norton.

Activity #36: Greenhouse Globe

Botkin, D. B., M. F. Caswell, J. E. Estes, & A. A. Orio. 1989. *Changing the Global Environment: Perspectives on Human Involvement*. Boston: Academic.

Golden, R. (ed.). 1989 (Spring). *Greenhouse Gas-ette*. Oakland, CA: Climate Protection Institute.

Ohio Sea Grant Education Program. *Global Change in the Great Lakes: Scenarios*. Columbus, OH: Ohio State University.

Schneider, S. H. 1989. *Global Warming: Are We Entering the Greenhouse Century?* San Francisco: Sierra Club.

Activity #37: Kirtland's Warbler

Botkin, Daniel B., D. A. Woodby, & R. A. Nisbet. 1991. Kirtland's warbler habitats: A possible early indicator of climatic warming. *Biological Conservation, 56*: 63-78.

Byelich, J., M. C. DeCapita, G. W. Irvine, R. E. Radtke, N. I. Johnson, H. Mayfield, & W. J. Mahalak. 1985. *Kirtland's warbler recovery plan*. Rockville, MD: U.S. Fish and Wildlife Service, 78 pp.

Zimmerman, D.R. 1975 (May). Panic in the pines. *Audubon, 77*: 88-91.

Activity #38: Arctic Warming

Oechel, W. C. 1994. Personal communication.

Oechel, W. C., S. J. Hastings, G. Vourlitis, M. Jenkins, G. Riechers, & N. Grulke. 1993 (Feb. 11). Recent Change of Arctic Tundra Ecosystems from a Net Carbon Dioxide Sink to a Source. *Nature, 361*: 520-523.

Activity #39: Acid Rain

Hilchey, T. 1993 (Sept. 7). Government Survey Finds Decline in a Building Block of Acid Rain. *New York Times*, C4.

Mohnen, V. A. 1988 (Aug.). The Challenge of Acid Rain. *Scientific American, 259*(2): 30-38.

Activity #40: House Trees

Botkin, D. B., & C. E. Beveridge. 1992. *Vegetation in SemiArid Cities: An Introduction to the Uses and Benefits of Vegetation in Urban Environments of Southern California*, Research Paper No. 921001. Santa Barbara, CA: The Center for the Study of the Environment.

Heisler, G. M. 1986. Energy Savings with Trees. *Journal of Arboriculture, 12*(5). In *Cooling Our Communities: A Guidebook on Tree Planting and Light-Colored Surfacing*. U. S. EPA Office of Policy Analysis, U. S. Superintendent of Documents, ISBN 0-16-036034-X, Washington, DC.

Activity #41: Green Streets

Dreistadt, S.H., D. L. Dahlston, & G. W. Frankie. 1990. Urban forests and insect ecology. *Bioscience, 40*(3): 192-197.

Moffat, A.S. 1987 (July). Killing streets. *Horticulture, 65*(7): 56-61.

Moll, G., P. Rodbell, B. Skiera, J. Urban, G. Mann, & R. Harris. 1991 (April/May). Planting new life in the city. *Urban Forests, 11*(2): 10-17.

Activity #42: Plastics by the Numbers

Porro, J.D., & C. Mueller. 1993. *The Plastic Primer*. Washington, DC: League of Women Voters Education Fund.

Activity #43: Trash Collection

Murphy, P. 1993. *The Garbage Primer*. Washington, DC: League of Women
    Voters Education Fund.

Activity #44: Bottle Composting

Jones, L. L. C. 1992 (Oct.). Strike It Rich with Classroom Compost. *The
    American Biology Teacher, 54* (7): 420-424.
Sekscienski, G. 1992 (July/Aug.). Speaking of Composting. *EPA Journal, 18*(3):
    14.

Activity #45: Risk Assessment

Belton, T., R. Roundy, & N. Weinstein. 1986 (Nov.). Managing the Risks of
    Toxic Exposure. *Environment, 28*(9): 19-37.
Gunter, M.E. 1994. Asbestos as a Metaphor for Teaching Risk Perception.
    Journal of Geological Education, 42(7): 17-24.
Morgan, M. G. 1993 (July). Risk Analysis and Management. *Scientific
    American, 269*(1): 32-41.
Patton, D. E. 1993 (Jan., Feb., Mar.). The ABCs of Risk Assessment. *EPA
    Journal, 19*(1): 10-14.
Walker, B., Jr. 1992(Jul./Aug.). Perspectives on Quantitative Risk Assessment.
    Journal of Environmental Health, 55(1):15-19.
Wartenberg, D., & C. Chess. 1992(Mar./Apr.). Risky Business: The Inexact Art
    of Hazard Assessment. *The Sciences*, 17-21.

Activity #46: Environmental Attitudes

Hueber, G. 1991 (Apr.) Americans Report High Levels of Environmental
    Concern, Activity. *The Gallup Poll Monthly*, 6-12.

NOTES

NOTES

NOTES

NOTES

NOTES

NOTES

NOTES

NOTES

NOTES

NOTES

NOTES